Facts of Matter and Light

Christian Maes

Facts of Matter and Light

Ten Physics Experiments
that Shaped Our Understanding of Nature

 Springer

Christian Maes
Physics and Astronomy
KU Leuven
Leuven, Belgium

ISBN 978-3-031-33333-0   ISBN 978-3-031-33334-7 (eBook)
https://doi.org/10.1007/978-3-031-33334-7

This Springer imprint is published by the registered company Springer Nature Switzerland AG
The registered company address is: Gewerbestrasse 11, 6330 Cham, Switzerland

Preface

While in school, I got profoundly interested in religion and in history. The reasons were probably similar for both subjects. First and foremost, I liked the stories but, like all youngsters, I was also looking for meaning and reason. Many of us like to understand things from the beginning. Having no real talent for religion though, and finding no relief in historical explanations either, I ended up thinking that mathematics might get me closer to truth. It has an attractive precision that is very tempting to a certain state of mind. Moreover, its language fit my simple abilities.

Mathematics does indeed make a wonderful spectacle of our world. It felt like riding a sturdy bike with an easy acceleration, using only pure thought as fuel. And it seemed that we could go everywhere with mathematics, it was so miraculously efficient. However, as teachers had warned me, its *casta* and the style of presentation started to bore me as well. In the end, physics was what I really wanted, as it appeared to be the science that dealt with the essential drama and the ultimate reality—you know, *to be or not to be*, that kind of thing. Mathematics started to become a kind of beloved gibberish with which I could speak about the universality of nature's laws, and express unifying principles from which a multitude of phenomena could be derived. There is indeed no better inspiration and line of attack for mathematics than good physics. That turned me into a mathematical, and then theoretical physicist, by psychological disposition, I suppose.

Naturally, this need not be the end of the story; things can get even more dirty … when physics gets really physical. After all, machinery and technology are awesome, as much a product of physics as they are useful for physics. A deep respect is owed to technological achievements, especially when based on simple ideas "that you can explain to the barmaid," as Rutherford would disrespectfully add. In this regard, I am not yet an experimental physicist, far from it; I'm still trying to learn. And it is a wonderful thing, learning how to make reliable experiments to question Nature, using today's wonderful technology which the development of science has made possible. Why this is not something everyone would like to do, getting ready for experiments, taking up the challenge of direct confrontation? Surely it must be fun and exciting to dialogue directly with Nature? If everything works out, asking questions and getting answers from the ultimate reality, what else did Moses dream of? Personally, I will probably never experience that glory. Yet I hope that the following chapters will encourage others, both young and old, to take an interest in scientific experiment …of the past and of the future.

The very idea of performing physics experiments is not self-evident. It never was, as history has shown, and for several reasons. Firstly, and quite trivially, thinking about an experiment obviously requires a minimal belief in the reality of an objective world out there. We need to care about matters of fact. Actually, we need to be interested in Nature, both what it is and how it is before we try to measure it. We have to be fascinated by this our Universe, its matter and energy in motion, which were there well before, and will be there well after our own temporal horizons on Earth. We often need to go beyond the measures of man as Vitruvius or Le Corbusier would have understood them.

In addition, there must be a strong conviction that evidence for what exists in the physical world can be provided reliably by empirical research. Controllable experiments are the key. We want reproducible evidence. Let us also hope that our senses and equipment do not fool us all the time, and combine that optimism with careful analysis. Evidence is often indirect, and ever more so as technology enters the lab in black boxes. We must build up experience to trust our experiments. That brings us to the relevant human qualities, such as patience and determination. But funding, pleasant collaboration, and the right equipment or infrastructure are essential as well.

Finally, over and above these practical issues and except for good luck, we must find the right questions. Physics is evidence-based, and experimental results often provide overwhelming evidence for the right line of thinking and for a particular theory. However, measuring ain't knowing. There are limits to Kamerlingh Onnes' phrase "Through measurement to knowledge." Whatever we believe we see or don't see depends on theory. In ever-growing circles, theory boldly proposes what is real and what is not, and experiment or observation approves or disapproves of it.

The present book is theory, but giving a central role to some of the most important experiments that have shaped our understanding of physics. Making a selection of physics experiments, be there 10 or 20 of them, is necessarily wildly arbitrary. Of course, choices can be discussed and rejected, and serious omissions noted. But there are mitigating circumstances. For one, I have tried to focus on historic experiments where there has been preparation and involvement by professional physicists. Somebody had to think about these experiments, set them up, and manipulate them in a reproducible way in the lab. I am not listing applications or mere observations, restricting myself to specific original achievements and fundamental physics. Obviously, the boundary between "experiment" and what is "only observation" remains vague and many intersections exist. I will not be dogmatic: during and after experiment come observation, reflection, and theorization. In fact, my purpose is not to make "the definitive list" of the most famous, groundbreaking, or revolutionary physics experiments. Here, such a list has one goal: to introduce interesting and important physics where the experiments stand at the beginning. Physics is an empirical science, in close interaction with the views or theories we have about nature. That interaction is what has made it so famously pleasant, and what is worth writing and reading about.

The plan is as follows: we start in the first chapter with a presentation of the items in our list of the most important physics experiments. This sets the stage—dealing with the first question of what this book is about. It is very open to debate whether the list is complete, whether it fails in places and neglects in others, or whether it overrates some experiments and underrates others. I only give a short *laudatio* there. In the following chapters, matters are clear: each type of experiment gets one chapter. We give more details and set them in the context of physical theories and views. We take the time to inspect the wider picture and develop ideas, even adding some mathematical touches here and there. In a good way, each chapter thus sets out to be a justification, explaining why the experiment is so important. It necessarily relates to the criteria I have used to produce the list. There have been two such touchstones.

A first motivation for highlighting an experiment is simply that it ought to represent a great discovery. It may be a confirmation of some thought or theory, or it may be telling us something deep and wonderful about how Nature is. These are revelations. A good example is Rutherford's experiment (1911) establishing the structure of the atom. It followed the Geiger and Marsden experiment (1909), in which it became clear that the plum pudding description of the atom put forward by Thomson was wrong. These scattering experiments gave us a picture of the atom and an estimate of the size of its nucleus. That discovery was foundational for the whole of twentieth century physics and chemistry.

The second motivation is to include experiments that have brought about changes in the very way we speak about Nature. The experiments on free fall, and in particular, the equivalence of gravitational and inertial mass, were so important for kick-starting classical mechanics and indeed general relativity, breaking almost completely with Aristotelian physics. We all know about free fall; it is easy enough to witness, but the controlled experiments set up by Galileo opened the way to a new science, i.e., science as we know it today.

There are other examples, and each time, I will come back to the motivation that got them onto the list. Apart from the two main criteria introduced above, there are of course also aspects of historical importance, impact, or beauty. All counts, and in the end, personal preferences matter as well. Speaking of which, there is a final chapter, just before the epilogue, where more speculation is offered about where to expect the most groundbreaking discoveries and game-changers in the near future. It also gives an opportunity to broaden the scope and reach out to other disciplines, including biology or psychology. All in all, modern physics remains open and forever renews its questions, often like a phoenix rising from the ashes of a previously held view. So we should not dwell too long on the experiments of the past; and here is a good reason not to write too much about them.

Few books on physics today emphasize experimental work. What is worse, too many popularizations in physics, while claiming to deal with physics and to address a general readership, are exceedingly speculative, theoretical, and more obscure than they are deep. Experiments are sometimes even pushed into the background, and made invisible, as if theory were the alpha and the omega of physics. There is nothing wrong with abstraction, on the contrary, but education in physics need not and should not center around mathematical and theoretical physics. To get a good start, we would do better to gain some familiarity with the phenomena of Nature and with the idea and practice of doing experiments. There, the great performances of the past still provide excellent stepping stones. We may take the words of Maxwell to hold for theoretical and experimental scientists alike:

> The first processes, therefore, in the effectual studies of the sciences, must be ones of simplification and reduction of the results of previous investigations to a form in which the mind can grasp them.

—J. C. Maxwell, in *On Faraday's lines of force.*

The present book is written primarily with Master students in science or philosophy in mind as potential readers. Students of physics, in particular, will have heard most of what is being discussed and they will probably even have studied large parts of the material. Now they have to make choices. But let us not underestimate the pleasure in reading something we already know and understand, while contemplating farther horizons.

Granada, Spain   Christian Maes
March 2022

Acknowledgments

The four necessary human relationships: to love; to be loved; to be a teacher; to be a pupil.

—W. H. Auden, 1936.

It is a pleasure to thank Urna Basu, Pierre de Buyl, Enrico Carlon, Andrea Gambassi, Irene Maes, and Ward Struyve for suggesting changes or additions. Special thanks to Kasper Meerts for input, to Aernout van Enter for careful reading, and editing, and also to Faezeh Khodabandehlou for much help with the figures and index pages.

I also thank the Springer team and especially Angela Lahee for their encouragement.

As the reader will hopefully notice from what is to come, the motivation behind much of the text is simply a love of this subject, and the hope of transmitting my enthusiasm for understanding Nature on the basis of empirical findings. Obviously, my work is very limited in that respect and deep gratitude must go to my teachers.

Besides all the thinking and doing, the streamlining of research, and the organization of knowledge transfer to future generations and more distant students, it will always be the love of what supersedes us that brings lasting overall vitality to our subject.

Langdorp, Belgium   Christian Maes
December 2022

Contents

Chapter 1
The Winners Are …

Maybe here, before reading on, you should stop for a moment and ask yourself what *you* would take to be the 10 or 20 most important or greatest discoveries in experimental physics. Do not include mere observations, no matter how important, such as the discovery of the expansion of the universe or of structure in the cosmic background radiation. Do not even include the development of new experimental tools or methods, no matter how essential they may be. Let us limit ourselves. Just think about a lab and people setting things up in a precise way, to question Nature and learn in an active way about the physical world. What would you include in your list of the most famous, the most important, and the most groundbreaking experiments in physics?

As a matter of fact, I have found it interesting to put that question to students and colleagues: "What would be your list?" To my surprise, the things that got mentioned were often those that had been in the news most recently, and people spontaneously referred to pictures of black holes, gravitational waves, the Brout–Englert–Higgs particle, and so on. Some answered CERN, for short, not even adding explanation. I concluded that today's PR and outreach services associated with Big Science do a good job. They are worth their money. However, I must apologize to all those people: here I have ended up collecting mostly older and often technically less sophisticated things. Nevertheless, most of my colleagues converged upon topics and experiments that do indeed greatly overlap with the list below. In that sense, the choice is not so original after all; but then not much else could be expected, I would say.

Inevitably, however, there are serious omissions. To name a few, I have in mind the discovery of the Josephson effect (1962) in which an electric current can flow with no applied voltage through a thin insulating layer separating two superconductors, and the thermoelectric (Seebeck, Peltier, Thomson, and Ettingshausen) effects of the nineteenth century where energy-momentum, charge and particle currents interfere. Or, going way back, we can think of the hydrostatic experiments by Archimedes (287–212 BC) and the experimental development of optics, e.g. by Ibn al-Haytham (965—ca. 1041). Surely, other names and experiments could be added.

© The Author(s), under exclusive license to Springer Nature Switzerland AG 2023
C. Maes, *Facts of Matter and Light*,
https://doi.org/10.1007/978-3-031-33334-7_1

Here, then, are my 10 winners for the greatest experiments. I include a short motivation for each item so that they don't stand completely in isolation. This will also avoid possible misunderstandings—it will clarify what I have in mind. The order in the list is more or less arbitrary, except perhaps for chronology, but even that is not always justifiable. Just for fun, I tried to restrict the list to 10 (types of) experiments. In fact, as we shall see, even though each subject is fairly specific, some items do involve a number of different experiments. I can easily imagine rewriting the same list as composed of 12 or maybe even 20 different experiments. Yet, in honor of the Arabic counting system, here are 10:

1. **Free fall: the highways of the cosmos.** From the experiments of Galileo (from approximately 1600) to the synthesis by Newton (about 100 years later) and the revolutionary ideas of Einstein (some 300 years later), free fall has taken a central role in our understanding of mechanics and of the structure of spacetime. With Galileo, controlled free fall experiments confirmed that physics could be an empirical science. The older Aristotelian physics gave no way to understand free fall. Newton unified the free fall of apples with the motion of the Moon around the Earth, and Einstein used the equality of inertial mass and heavy mass to rethink the very nature of spacetime.

2. **Electromagnetism and optics unified: the dynamics of fields.** Electromagnetism achieved its first unification in the nineteenth century. Thanks to Faraday's experiments, the idea of a nonmaterial physical field was born. The experiments by Oersted (1820), Henry (1886), and, most importantly, Faraday (1839) gave the dynamical input to Maxwell's theory of electromagnetism and light (1861–62). From there on, large parts of physics, including gravity, would be formulated as field theories.

3. **Brownian motion: atoms exist, as revealed by fluctuations.** Since ancient times, and as revived in nineteenth century chemistry, there has always existed an idea that matter might be corpuscular. However, the hypothesis that there are indeed atoms took a long time to demonstrate. The experiments by Perrin (1908–1911) finally gave evidence of that atomic and molecular reality: it became possible to count the number of particles in a mole. As foreseen and convincingly explained by Boltzmann, this allowed a bridge (called entropy by Clausius) between the microscopic laws and macroscopic behavior. Thermodynamics got its foundation in statistical mechanics. More generally, statistical reasoning revolutionized physics through the power of large numbers (of photons, atoms, and molecules) and the idea of fluctuations obtained a physical grounding.

4. **Quantization: radiation, charge, and spin.** Atomism gave us a corpuscular view of matter and the discovery of the electron quantized charge. Calorimetric experiments and, later, spectral lines for atomic transitions such as the Lyman–Balmer–Paschen series in hydrogen, the black-body radiation experiments (1899), the photoelectric effect (Hertz, 1887) and Compton scattering (1923) were the most important stepping stones in demonstrating the corpuscular or quantized nature of light and radiation coupled to matter. Much of quantum physics found its inspiration in the experimental study of thermal radiation and

of the interaction of light with matter. And sometimes, what was thought to be a quantization of a classical concept, such as the direction of the orbital angular momentum, turned out to be a fundamentally new quantum property. I am referring here to spin and the paradigmatic Stern–Gerlach experiment which set the ball rolling.

5. **Wave-like behavior of light and matter**. The Young experiment (1802) showed that light is wave-like and the Thomson (1927) and Davisson–Germer experiments (1927) showed the same for electrons via diffraction in crystals. *A corner of the great veil was lifted*[1]: motion of sufficiently small particles is guided by waves. This realized the ideas of de Broglie and Schrödinger. Diffraction and interference pointed the way to a new (quantum) mechanics, expressed in terms of wave functions on configuration space.

6. **Structure of atoms and nuclei: scattering and fission**. Light scattering was shown to reveal dynamical properties of fluids in the Tyndall (1869–1869) and the Raman (1922, 1928) experiments. From the Rutherford scattering experiment (1911), identifying the nucleus of the atom, to the discovery of elementary particles confirming the standard model (period 1970–2010), scattering experiments have been our main microscope for examining the subatomic world. The production of the first nuclear chain reaction (1942), led by Fermi, was also the first controlled transmutation of elements, unleashing atomic power both for human use and for destruction.

7. **Invariance of proper time: the measure of light**. The mechanics of light formed a dark cloud over the physics landscape at the end of the nineteenth century. Did light need a medium, a so-called luminiferous æther, in order to propagate? The Michelson–Morley experiment (1887) was a crucial step on the way to a new mechanics, in which spacetime would reveal new symmetries. Minkowski's description of spacetime was a direct mathematical formulation for taking the Michelson–Morley experiment to its natural conclusion. The experimental basis for relativity started from there, while interferometry went on to provide an ever more accurate technique for investigating the properties of spacetime.

8. **Vacuum polarization: dynamical activity of the vacuum**. *You'll never walk alone* ... Zero-point energy and vacuum fluctuations are part of this, our quantum world. The Lamb–Retherford experiment (1947) made us look differently at the vacuum. Nothingness would never be the same again. Modern quantum electrodynamics took off with those microwave radiation experiments, establishing the Lamb shift and introducing a crucial role for renormalization ideas in quantum field theory. A fluctuating vacuum as background for fundamental processes leads to the observed Casimir effect (1948–97), while the origin of the van der Waals force and the London dispersion forces are conceptually similar.

---

[1] *Er hat einen Zipfel des grossen Schleiers gelüftet*, from a letter of Einstein to Langevin on 16 December 1924. The same day, Einstein wrote to Lorentz that *this is a first, weak ray to cast light on this worst of all our physical puzzles*. That same month, Einstein completed his manuscript "Quantum theory of the monatomic ideal gas" predicting Bose-Einstein condensation (Nobel prize 2001).

9. **Emergence: phase transitions**. The exploration of a new world, both unexpected and rich, resulting in various collective or many-body phenomena, made the most sensational discoveries in condensed matter theory: dynamical symmetry breaking, phase transitions and critical phenomena. From Andrews' classic experiment (1863) on the liquefaction of gases, to the discoveries of superconductivity (Kamerlingh Onnes, 1911a, 1911b) and superfluidity (Kapitza et al., 1938), and with the creation of a pure Bose–Einstein condensate (in 1995 by Cornell, Wieman, and collaborators), explorations of the equilibrium phase diagram of many-body systems left the classical domain and entered the arena of quantum physics. Low-temperature physics became a hot topic.

10. **Spooky action at a distance: Nature is nonlocal**. And of all things the weirdest is that entanglement operates over arbitrarily large distances, enabling action at a distance. The Bell inequality is indeed violated as first seen from experimental work by Freedman & Clauser (1972) and Aspect et al. (1982), and by later worldwide confirmations in the form of Bell tests. Related thought-experiments on the completeness of quantum mechanics launched a second revolution, where individual light and elementary particles can be individually manipulated. Reflection on quantum foundations led to quantum technology, as in the experiments of Zeilinger (such as quantum teleportation in 1997), providing jump-starts for new applications in quantum information and communication. Theoretically, the nonlocal aspect of reality remains poorly understood, but entanglement and the combination of quantum optics with condensed matter physics are making new waves in the exploration of many-body quantum phenomena.

The above *laudationes* indicate the content of the chapters that follow. The experiments will be discussed in more detail there, but here is the shortlist (with all its shortcomings):

1. Galileo's experiments on free fall
2. Faraday's induction experiment
3. Perrin's experiment on Brownian motion
4. Blackbody radiation experiments
5. Young's double-slit experiments
6. Rutherford's scattering experiment
7. The Michelson–Morley experiment
8. The Lamb–Retherford experiment
9. Kamerlingh Onnes' discovery of superconductivity
10. Bell test experiments

Again, the above is (only) a list of keywords. In each case, several experiments (and theory) hang together. After each discussion, we will look back and we will look forward. The final chapter of the book will try to emphasize how the main waves of progress in physics have been made by absorbing and unifying different fields of application and by exploring different levels of empirical observation. In this respect, we can suppose that the future of physics and future experiments will not be very different. However, let us hope that future lists will no longer reflect a "man's world[2]" of physics as flagrantly is the case for the present list.

Before we reach those final points, however, we shall start with very standard material, especially in the first two chapters on classical mechanics and electromagnetism. Not until Chaps. 4 and 5 will we enter the twentieth century with the mutually related ideas of fluctuation and quantization.

---

[2] "Biblically chauvinistic" as *Rolling Stone* magazine characterized the lyrics by Betty Jean Newsome of the song *It's a Man's Man's Man's World*.

Chapter 2
Free Fall

A central problem in physics is to understand how stuff moves. The study of thrown and falling objects has considerably advanced our knowledge of spacetime and the way matter traces its path. The very beginning of classical mechanics in fact goes back to just such experiments. The basic question of free fall is to compute changes in position, while leaving things to themselves and gravity, after giving them an initial position and velocity.

It is only really from the seventeenth century onward that we learn from studies of free fall that mechanics is possible. Before that time, the overwhelming range of different types of motion left little hope of finding a unifying description of what appeared to be a chaotic array of different phenomena. How to see harmony in a wasteland?

One excellent choice was to concentrate more modestly on just one type of motion, namely free fall. This led to the discovery of a systematic and reproducible approach, whence law-like behavior could at last be unveiled by careful experiment and observation. Free fall did indeed seem to involve an extra simplification: we are all affected in the same way under gravity, as decided by our mass alone (at least in vacuum). I.e., bodies on Earth all fall with the same acceleration when unhindered. And motion in our sublunary sphere is not essentially different from the motion of superlunary celestial bodies. We can speak about a universal theory of gravitation. From there, it seemed less outrageous to think that motion might show universal features; that mathematics and careful experiment might combine in a powerful way to provide us with an understanding.

© The Author(s), under exclusive license to Springer Nature Switzerland AG 2023
C. Maes, *Facts of Matter and Light*,
https://doi.org/10.1007/978-3-031-33334-7_2

The possibility of mechanics and its universality meant drastic departures from the Aristotelian view of motion. That was true in many ways.[1] For one, the very method of investigating nature had changed. The study of free fall introduced the important notion of a controlled experiment into the study of Nature. Just as we can learn to cook a dish in a reproducible way, maybe given a recipe, experiments can become trustworthy and reproducible. We can repeat them and check the results anywhere in the world. That empirical method stands in stark contrast to the more conceptual technique of inquiry, more at home in philosophy, which has traditionally (but often baselessly, Russo (2004)) been attributed to the Ancient Greeks. We hold therefore that empirical science started by dropping and throwing things. From there, classical mechanics took off.

As a matter of fact, almost all of classical mechanics and the theory of dynamical systems was inspired by celestial mechanics, where gravity is the main (or only) force. It shaped the very idea of force and interaction, which remained unchallenged for about 200 years. It is therefore quite fitting that the revolution of 1907–1917, which introduced the general theory of relativity and with it a new framework for physical cosmology, started once again from reflections on free fall. It would bring together the concepts of spacetime and matter, and the motion of matter, which in Newton's time were still considered independent.

2.1 Equality of Gravitational and Inertial Mass

Prior to the sixteenth century, at least in the Western world, if people thought of falling bodies at all, they thought of it in terms of their everyday experience: things would fly in straight lines, first keeping their initial direction, lose speed "somehow," then drop vertically to the ground. It was thought that the speed of a falling body would be proportional to its weight, a belief that may have come from watching light things whirl and swirl in a fluttering movement down to the ground. This is not so unreasonable and we can easily believe it from experience, especially in the autumn. It was perhaps not accepted by all, but many believed that a 5 kg object would fall 5 times faster through the same medium than an otherwise identical 1 kg object. Some scholars certainly referred to Aristotle (350 BC), who did indeed discuss falling objects in his book VII of *Physics*. The idea of what determined free fall, what was natural, and for what purpose played an important role in people's world view, as expressed for example by Thomas Aquinas (1225–1274) in his *Commentary* of 1268–1269. And this was the main stream of thought in the Arab and Western world until the time came for people to look more carefully at free fall.

---

[1] Not only Galileo Galilei (1564–1642), but also Renaissance atomists such as Isaac Beeckman (1588–1637) and Sébastien Basson (ca. 1573–1625) were often very critical toward Aristotle, as e.g., explicitly in the title "Twelve books of natural philosophy against Aristotle" of Basson's treatise from 1621. It is little unfair though that Aristotle was used as a sitting duck. The paradox is that Aristotle was in many ways more empirically focused than the atomists and natural philosophers of Ancient Greece.

Who was first and who did what in describing free fall is not so clear, however, and neither is it so very important; see Dijksterhuis (1950). Among the names are Giambattista Benedetti (1530–1590), who clearly suggested (1553) that bodies of equal weight would touch the ground at the same time when dropped simultaneously. Well documented also is the experiment by Jan Cornets de Groot (1554–1640) and Simon Stevin (1548–1620) in the second half of the sixteenth century. They climbed the tower of the "Nieuwe Kerk" (a church) in Delft (the city of the much loved seventeenth century painter Johannes Vermeer (1632–1675) in what is now called the Netherlands). They dropped two lead balls of different masses from a height of approximately 30 m and thus saw proof that heavy and light objects fall at the same speed:

> Take (as H. Ian Cornets the Great, man of learning and most diligent student of the mysteries of nature, and I have done) two lead balls, one ten times larger and heavier than the other. Let them fall together from a height of 30 feet, to fall on a board or something that makes enough noise. It will appear that the lightest one is not ten times longer on its way, but that they reach the board so equally that both sounds seem to be from the same knock. The same happens with two bodies of equal size, one of which is ten times heavier than the other, which means that Aristotle's law of proportionality is false.[2]
> — Simon Stevin, in *De Beghinselen der Weeghconst*, 1586 (translation by the author).

That the two lead balls, one 10 times bigger and heavier then the other, reached the ground at the same time, they judged from the sound the balls made, hearing a single bang when they finally came to rest.

And there is also the tale, maybe apocryphal, that Galileo in 1589–92 dropped two objects of unequal mass from the leaning tower of Pisa. He describes the experiment as follows:

> But I, Simplicio, who have made the test, assure you that a canonball that weighs one hundred pounds (or two hundred, or even more) does not anticipate by even one span the arrival on the ground of a musket ball of no more than half, both coming from the height of two hundred brachia.
> — Galileo Galilei, in *Discourses and Mathematical Demonstrations Relating to Two New Sciences*, 1638.

At any rate, Galileo already discussed free fall experiments extensively in his Dialoghi (1632). Other people such as Francesco Grimaldi (1618–1643) and Giovanni Riccioli (1598–1671) investigated free fall between 1640 and 1650, confirming that the distance fallen goes as the square of the time. They also made a calculation of the gravitational constant by recording the oscillations of an accurate pendulum. Despite this, Riccioli seemed to conclude that heavy bodies fall faster. In Flanders,

---

[2] Laet nemen (soo den hoochgheleerden H. IAN CORNETS DE GROOT vlietichste ondersoucker der Naturens verborghentheden, ende ick ghedaen hebben) twee loyen clooten d'een thienmael grooter en swaerder als d'ander, die laet t'samen vallen van 30 voeten hooch, op een bart oft yet daer sy merckelick gheluyt tegen gheven, ende sal blijcken, dat de lichste gheen thienmael langher op wech en blijft dan de swaerste, maer datse t'samen so ghelijck opt bart vallen, dat haer beyde gheluyden een selve clop schijnt te wesen. S'ghelijcx bevint hem daetlick oock also, met twee evegroote lichamen in thienvoudighe reden der swaerheyt, daerom Aristoteles voornomde everedenheyt is onrecht.

Juan Caramuel de Lobkowitz (1606–1682) also used a small pendulum to measure time. However, none of these appear to have taken into account differences between falling in air and falling in a vacuum, as Galileo and Newton did later. The most recent notable iteration of this experiment was carried out in 1971, by David Scott. Although not very accurate, it was certainly the most dramatic performance as it was done on the Moon as part of the Apollo 15 mission. At the end of the last Moon walk, Scott held a hammer and a feather in front of him in view of a television camera, and let them go at the same time. Unsurprisingly, in the ultra-high vacuum of space, they hit the lunar surface at the same time.

In the formalism of Isaac Newton (1643–1727), where gravity is a property created by objects with gravitational mass and experienced by objects with inertial mass, a little reflection shows that the universality of free fall is equivalent to the equality of these two types of mass for every substance. (To be pedantic, it only requires their proportionality.) In accord with its fundamental nature, this principle is affirmed by Newton just after Definition 1 in his *Principia*, the definition of mass.

> It is this quantity that I mean hereafter everywhere under the name of body or mass. And the same is known by the weight of each body; for it is proportional to the weight, as I have found by experiments on pendulums, very accurately made, which shall be shewn hereafter.
> — Isaac Newton, in the *Principia*, 1687.

Newton knew that this was not true for forces in general (such as magnetism), implying that the universality of free fall was almost an accidental fact, not something that could be derived from first principles, but which in fact had to be empirically verified. In his experiments, Newton fashioned two pendulums out of wooden boxes filled with different materials, equal in weight and size to avoid any differences in air resistance. Putting them side by side, he observed them to swing equally for a long while, confirming the equality for gold, silver, lead, glass, sand, common salt, wood, and wheat. Curiously, he did not include mercury, a heavy substance that was otherwise very present in his alchemical work.[3] Compared to the difficulty in determining when exactly two falling objects hit the ground, these experiments were vast improvements in accuracy.

> And by these experiments, on bodies of the same weight, I could have discovered a difference of matter less than the thousandth part of the whole.
> — Isaac Newton.

In later experiments, any difference between inertial and gravitational mass was further shown by Friedrich Wilhelm Bessel (1784–1846) to be less than one part in 50000. As an anecdote, Bessel also included meteorites in his list of substances, in an attempt to find a distinction between earthly and non-earthly substances regarding their attraction to the Earth. He remarked that it would be interesting to keep checking this assumption with the ever-increasing precision provided by the new instruments of each future generation....

---

[3] Perhaps the problem lay in this heavy substance being liquid. Turbulent fluctuations would have proved unavoidable and detrimental to accuracy.

The same experiments on free fall do indeed continue today in modern forms and to ever greater accuracy, in what are called Eötvös experiments. This started around 1885 when Loránd Eötvös (1848–1917) used a torsion balance, and it has grown to tests for checking the equivalence for the free fall of elementary particles. There are no doubt still things to discover there. For example, we would like to observe the gravitational interaction of antimatter with matter or antimatter. Most likely, gravity attracts both matter and antimatter at the same rate that matter attracts matter, but it would be very nice to obtain experimental confirmation.

2.2 Galileo's Experiments on Free Fall

Given the speed at which balls fall from a tower of sufficient height, not much information about the true and quantitative nature of free fall can be extracted from such an experiment. Galileo opted for something more sophisticated and more controllable that would open avenues to the wider field of mechanics. Galileo did experiments with bodies rolling down ramps, as in Fig. 2.1. Here, the motion was slow enough to be able to reliably measure the relevant time intervals with water clocks and his own pulse. He could also repeat them as often as needed, noting that, repeating "a full hundred times," he could achieve "an accuracy such that the deviation between two observations never exceeded one-tenth of a pulse beat." In particular, he used balls rolling on inclined planes to show that different weights fall with the same acceleration. Regardless of the angle at which the planes were tilted, the final height the ball reached on the opposite side almost always equalled the initial height.

Now comes the idea of inertia (see Fig. 2.1). Reducing the slope of the opposite incline, the ball rolled a further distance to reach its original height. Fixing the elevation at an angle close to zero degrees, the ball would roll almost forever, as though nevertheless trying to reach the original height. And if the opposing plane was not tilted at all (that is, if it were oriented along the horizontal), then …an object in motion would continue, if possible, to go right around the Earth. For Galileo,

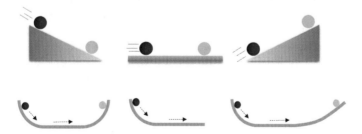

Fig. 2.1 Experimenting with motion on inclined planes was a better controlled way to reveal the mechanics of free fall. It played a major role in Galileo's pioneering contributions to classical mechanics

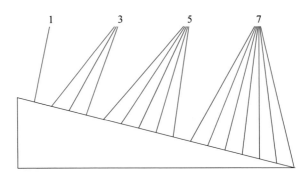

Fig. 2.2 For motion on an inclined plane, the particle's mass does not affect the acceleration a. Only the angle θ of the slope is relevant, through the relation $a = g \sin \theta$. Galileo's law of odd numbers states that the distances traveled by a free-falling body in equal successive periods of time are proportional to the succession of odd numbers 1, 3, 5, 7, ... Courtesy of Kasper Meerts

motion was as natural as rest, and moving objects would eventually stop because of a force called friction.

It is important that these experiments were accompanied by quantitative measurements (Fig. 2.2). Galileo discovered that the distance fallen was proportional to the square of the time it had been falling. Many of the results were already summarized in *De Motu Antiquiora*, an unpublished manuscript on the motion of falling bodies that dates from the period 1589–92. His final "testament" (Discorsi, 1638) is obviously more detailed.

Experimenting, and in particular experimenting with free fall, opened the skies, almost literally. Galileo's work seriously broke with tradition at that time in Western Europe Dijksterhuis (1950). He believed henceforth that there could be no perfect, unchanging sphere around the center of the universe,indexuniverse as Aristotle had claimed. Feelings about how the universe should look had to change drastically (again).

Galileo was not the first to subscribe to these changes, however. He adopted Copernicus's heliocentric hypothesis and thought of planet Earth very much as just another planet. In fact, his discoveries with the telescope helped revolutionize astronomy and accept the Copernican system. In this respect, observation via technical means, experimentation, and quantification were no longer seen as *sinful* or in some way opposed to Nature. That brings about a unification of mechanics (*techné*) and physics. He thus contributed in an essential way to the development of the scientific method, where systematic experiment combines with mathematical (geometric) and quantitative analysis, and teleological arguments are rejected.

Naturally, it would be a major exaggeration to consider Galileo as the only person or indeed the first involved in establishing modern experimental science. Hellenic science also had plenty of examples of experiments and systematic observations, Russo (2004); Kuhn (1957). Furthermore, it is not true that everything began in Western Europe. For example, the classical works of al-Biruni (600 years earlier) and al-Khazini also stand at the beginning of the application of experimental methods in medieval science, continuing from the work of Archimedes. Nevertheless, Galileo played a crucial role in the development of classical mechanics and the birth of science, before Newton eventually took over.

2.3   Newton's Gravity

2.3.1   *Looking up at the Sky*

Science usually follows a reductionist programme. This implies unification, connecting a multitude of physical phenomena to fewer, more elementary considerations. Nature is not just a collection of facts for the scientist. Learning about Nature requires the discovery of underlying regularities, laws if you wish, harmonies if you prefer. In the case of free fall, this comes with a story, or rather a legend, most likely, the legend of the falling apple in the garden at the Woolsthorpe Manor, the birth place and family home of Isaac Newton.

Isaac had left Cambridge, where he was a student at Trinity College, to escape the plague. This was 1665, which saw a serious outbreak of the bubonic plague in England. And so we picture him in the garden at his parents' house in the English countryside, resting, enjoying a happy summer's day. Then an apple falls from the tree under which he is sitting and, so the story goes, Newton suddenly realized at that very moment how each and every object behaves in the same way under free fall. This gave him the inspiration to formulate the universal law of gravitation. Newton later relayed the apple story to William Stukeley, antiquarian, who included it in his "Memoir of Sir Isaac Newton's Life," a book already published in 1752.

Looking up at the sky from time immemorial, people had long wondered about regularities and looked out for special occurrences[4] Just as today we look out for rain and consequences of climate change, so people have always expected signs from above, signaling good times for harvests or perhaps the right moment for rituals, revealing the advent of a new life, or of disaster. As the constellations shift over the year, planets wander between them, while the Moon appears to stay close by, watching us from all angles. Indeed, we may come to wonder what decides change. And in that case, we have a friend in Newton, for his publication, the *Principia*, or "Mathematical Principles of Natural Philosophy" is the most eminent book of changes, the Western European *I Ching*, telling us how to think about change and how to compute trajectories as time unfolds.

---

[4] In contrast with some interpretations, one can easily imagine here the stimulating influence of religion and superstition for the development of mathematics and natural sciences more generally. Newton is probably a good example, but also Copernicus or Galileo have been supported and often positively influenced by organized religion. As a more trivial case, also today and more than once, a student starts studying physics driven by an honest desire to understand books and videos that are more obscure than they are deep.

Fig. 2.3 Landscape with the
Fall of Icarus (1558), by
Pieter Breugel the Elder

About suffering they were never wrong,
The old Masters: how well they understood
Its human position: how it takes place
While someone else is eating or opening a window or just walking dully along;
How, when the aged are reverently, passionately waiting
For the miraculous birth, there always must be
Children who did not specially want it to happen, skating
On a pond at the edge of the wood:
They never forgot
That even the dreadful martyrdom must run its course
Anyhow in a corner, some untidy spot
Where the dogs go on with their doggy life and the torturer's horse
Scratches its innocent behind on a tree.

In Breughel's Icarus, for instance: how everything turns away
Quite leisurely from the disaster; the ploughman may
Have heard the splash, the forsaken cry,
But for him it was not an important failure; the sun shone
As it had to on the white legs disappearing into the green
Water, and the expensive delicate ship that must have seen
Something amazing, a boy falling out of the sky,
Had somewhere to get to and sailed calmly on.
— Musée des Beaux Arts, by W.H. Auden (1907–1973), 1938 (Fig. 2.3).

Old texts suggest histories of long ago. The Bible also has them. As a keen theologian, Newton took an interest in reconstructing the chronology of these texts, detailing the rise and fall of the ancient kingdoms they chronicled. As in all ancient cultures, time managers were observers of the sky. So Newton got involved in trying to understand the mechanics of the Moon and the planets. And he did. As he would state himself, however, he could see further because he stood on the shoulders of giants, although he was especially sarcastic about Robert Hooke (1635–1703), England's Leonardo and main rival of Newton, who turned out to be "but of midling stature." There was certainly plenty of data available, as in the observations of Tycho Brahe (1546–1601) and Johannes Kepler (1571–1630). Above all, however, Newton credited Galileo directly for understanding the importance of inertia. He also mentioned that Galileo understood the principles of motion in a uniform gravitational field. Newton was Galileo's successor in more ways than one, but it is not quite true as is commonly claimed that Newton was born in the year Galileo died.[5]

2.3.2 Newton's Program

Isaac Newton was the culminating figure in the scientific revolution of the seventeenth century. His *Philosophiae Naturalis Principia Mathematica* (1687 for the first edition) is one of the most important works in the history of modern science. It sets out a program that rests on basically two laws; see Maudlin (2015) for a very readable introduction to the conceptual points.

First Law

> Every body perseveres in its state either of rest or of uniform motion in a straight line, except insofar as it is compelled to change its state by impressed forces.

This means "every" body, be it apples or moons. The force that maintains our own Moon in its orbit about the Earth is precisely the same force that causes an apple to fall from a tree. No peculiar natural motion is ascribed to any specific body, and in fact every body has an inherent tendency to *maintain* its state of motion.

Moreover, in stark contrast to Aristotle, there is no essential difference between motion on Earth or motion in the sky. That unification means there is no fundamental diversity in the nature of motion. Furthermore, there is no need to distinguish special regions in the universe. Rather, the arena of motion is *absolute space*. That space has a structure which gives meaning to the notion of rest and uniform motion in a straight line. Indeed, space and geometry are Euclidean. Absolute space is the 3-dimensional space E^3. We also need an absolute time dimension to gauge the passage of time, which always goes forward.

Note that Newton still makes a point of mentioning both "rest" and "uniform motion." That is because Newton believed that this 3-dimensional space exists at every moment of time, and moreover that the same points of space persist identically through time. For a body to be *resting* means that it remains at the same points of absolute space. Believing in absolute space is not so strange, but Newton even appealed to a thought experiment, now referred to as *Newton's bucket*, to try to argue for its existence.

> If a vessel, hung by a long cord, is so often turned about that the cord is strongly twisted, then filled with water, and held at rest together with the water; after, by the sudden action

---

[5] Galileo died on 8 January 1642 (Gregorian calendar) and Newton was born on 25 December 1642 (Julian calendar). When placed on the same calendar the two events fall in different years. Robert Newton, Isaac's father, died some 3 months before Isaac was born.

of another force, it is whirled about in the contrary way, and while the cord is untwisting itself, the vessel continues for some time this motion ; the surface of the water will at first be plain, as before the vessel began to move; but the vessel by gradually communicating its motion to the water, will make it begin sensibly to revolve, and recede by little and little, and ascend to the sides of the vessel, forming itself into a concave figure. This ascent of the water shows its endeavour to recede from the axis of its motion ; and the true and absolute circular motion of the water, which is here directly contrary to the relative, discovers itself, and may be measured by this endeavour. [...] And therefore, this endeavour does not depend upon any translation of the water in respect to ambient bodies, nor can true circular motion be defined by such translation. [...] but relative motions [...] are altogether destitute of any real effect. [...] It is indeed a matter of great difficulty to discover, and effectually to distinguish, the true motions of particular bodies from the apparent; because the parts of that immovable space in which these motions are performed, do by no means come under the observations of our senses.
— Isaac Newton; Principia, Book 1: Scholium.

Perhaps for us it is easier to imagine two spheres connected by a cord and rotating around their common center of gravity in an otherwise empty space. They too can be considered as rotating with respect to absolute space. There is absolute rotation, hence, absolute space and time.

In that sense Newton was an absolutist, much like Samuel Clarke (1675–1729), who, helped by Newton, had a now famous correspondence with Gottfried Leibniz (1646–1716), claiming that space was a substance or something substance-like (a "monadic property of God"). Leibniz thought that absolutism was false and defended relationalism, the view that spatial and temporal relations are more fundamental. Apparently, Leibniz was of the opinion that Newtonian physics was detrimental to natural theology. The correspondence (1715–1716) was cut short when Leibniz died.

Newton used the rotating bucket argument to demonstrate that rotational motion could not always be defined as the relative rotation of the body with respect to its immediate surroundings. In general, true motion and rest were to be defined by reference to absolute space. Alternatively, these experiments give us an operational way of defining "absolute rotation" without the need to specify what it was relative to."

Second Law

The first law formulates the state of motion of every body when there is no external force: it is either at rest or in uniform straight motion in and with respect to absolute space. That is, however, insufficient to account for our own world, where bodies are almost never seen in uniform motion. In the second law, Newton launches his program on forces:

Any change in motion is proportional to the impressed motive force, and occurs along a straight line in the direction of the impressed force.

By the first law, we know what is meant by a change in motion. From the second law, we say that, to keep a body say in uniform circular motion , there must always be a force (the centripetal force). The Moon is attracted to the Earth by such a force, namely gravity.

Here is how Newton's algorithm works. We use a one-dimensional notation for simplicity, where x is the position and v is the velocity of a point particle of mass m. We are interested in knowing the trajectory of the particle, i.e., the sequence of positions $x(0), x(\varepsilon), x(2\varepsilon), \ldots$, where ε denotes a small time unit. How do we obtain $x((n+1)\varepsilon)$ from $x(n\varepsilon)$? The answer uses the velocity, according to Newton (or the momentum to be more precise, but we need not stress that point here).

Let us start from the beginning, given $x(0) = x$, $v(0) = v$. At the next instant, we put

$$x(\varepsilon) = x + \varepsilon v , \qquad v(\varepsilon) = v + \varepsilon \Phi(x) ,$$

where the acceleration is written as a function (some function Φ) of x so that the algorithm can be iterated. At a general time, we need to know both the position $x(n\varepsilon)$ at time $n\varepsilon$ and also the velocity $v(n\varepsilon)$ at that time. Then we set

$$x\big((n+1)\varepsilon\big) = x(n\varepsilon) + \varepsilon v(n\varepsilon) .$$

The next problem is of course to find $v((n+1)\varepsilon)$, because we will need it to find $x((n+2)\varepsilon)$. Here comes Newton's central idea, which gets the algorithm to run. Newton states that $v((n+1)\varepsilon)$ can be found (again) from $x(n\varepsilon)$. The concept of force arises here:

$$v\big((n+1)\varepsilon\big) = v(n\varepsilon) + \varepsilon \, \Phi\big(x(n\varepsilon)\big) ,$$

where the function Φ only depends on the particle through its mass m and is otherwise "universal," so $\Phi(x) = F(x)/m$, where F is the force as decided by natural laws.

In this way, we can obtain the full trajectory of the particle; we need to know its initial position and velocity $(x(0), v(0))$, its mass m, and the force F exerted on it. That force F closes the algorithm by its dependence on position. Physics claims that this is how Nature works: the problem is to identify the forces. We work with second order differential equations, where the dynamics is determined by a force which has universal properties (the same for all bodies) and reflects the symmetries of space and time. That is a claim that can be checked empirically. In Newton's day, the only "known" force was gravity.

In passing, observe that the algorithm is deterministic and time-reversible.

2.4 Gravity

The law of gravitation presents an enormous mathematical unification of the various phenomena of falling bodies, from apples to moons and planets. It does not give a precise mechanism for gravity though, but replaces it with a unique mathematical formulation. As a matter of fact, that force of gravity is difficult to make sense of or even to interpret; it represents an action-at-a-distance: from the moment a mass appears, it instantaneously affects the whole of the rest of the universe.

> I have not as yet been able to discover the reason for these properties of gravity from phenomena, and I do not feign hypotheses. [...] That one body may act upon another at a distance through a vacuum without the mediation of anything else, by and through which their action and force may be conveyed from one another, is to me so great an absurdity that, I believe, no man who has in philosophic matters a competent faculty of thinking could ever fall into it.
> — Isaac Newton.

We had to wait another 200 years for the first solution. For the physical understanding, it was basically the work of a single person, Albert Einstein (1879–1955). Einstein's general theory of relativity (1915) is an extension (but also, a serious modification) of his special theory of relativity, where a central role is played by the (new) geometric nature of gravity. On the other hand, "spooky action at a distance" still appears to be part of Nature, as we will explain when discussing the notion of entanglement in Chap. 11.

In Newton's theory, inertial mass equals gravitational mass. We understood this above from the experimental fact that two balls of different mass, allowed to fall from rest at a certain height, hit the ground simultaneously. Since the force of gravity is proportional to the inertial mass and that same mass enters Newton's second law, the acceleration due to gravity is the same for all bodies. This statement is called the weak equivalence principle, and is left unexplained in Newtonian mechanics.

Einstein's idea was to turn the situation upside down. He proposed to think of free fall as a "natural" motion , entirely determined by the structure of spacetime alone (a structure which is influenced by its mass–energy content). Freely falling bodies feel no forces on them at all. That weak equivalence principle now comes for free: two bodies fall together with the same acceleration because they are in fact following the straight lines of free motion , the same straight trajectory through spacetime: there is no longer a "force of gravity."

The strong or Einstein equivalence principle states that the outcome of any experiment carried out in "free-fall" in a uniform gravitational field would have the same result as if it were carried out in an inertial laboratory in empty space, and any experiment carried out "at rest" in a uniform gravitational field will have the same result as one carried out in a uniformly accelerating laboratory in empty space. This strong equivalence principle is stronger than the weak equivalence principle because it has implications for *all sorts* of experiments. Think of light (where the concept of mass does not even enter into consideration and we can still choose to think of it as a wave or as particles); "freely falling light " follows the straight lines as dictated by the intrinsic geometry of spacetime, and the path of light will "bend" when it passes close to a star. Arthur Eddington's eclipse expedition of 1919 confirmed this very effect, and general relativity thereby became our leading account of gravity and has remained so ever since.

Chapter 3
Electromagnetic and Optical Unification

3.1 Electromagnetic Phenomena

> Who would not have been laughed at if he had said in 1800 that metals could be extracted from their ores by electricity or that portraits could be drawn by chemistry.
> — Michael Faraday, The Letters of Faraday and Schoenbein, 1836–1862.

It is easy to imagine how, if electromagnetic phenomena had been demonstrated in the centuries before Newton or Galilei, the poor experimentalists would have been brought up before the Inquisition. Electricity and magnetism might certainly look like the devil's work. The eighteenth and nineteenth centuries were safe enough in Western Europe, however. Experiments were conducted in high-society gatherings and indeed became a subject of regular investigation. Yet, the systematic scientific study and the theory of electromagnetism only started to develop in a serious manner from the nineteenth century. True, the electric and magnetic forces were known from way back, but it was only with Michael Faraday (1791–1867) that a new and more complete picture began to emerge. The synthesis was made by James Clerk Maxwell (1831–1879) in "his" four laws:

> Maxwell's theory is Maxwell's system of equations.
> — Heinrich Hertz, Electric waves (1893), 21.

Let us briefly recall some phenomena to illustrate those laws.

First Law: Electric Charge

We are most familiar with the electrical interaction. The most sensational example of electric discharge is lightning. The charge is stored in dark clouds or in the atmosphere more generally, and lightning occurs by the quasi-instantaneous release of around a gigajoule of energy, equivalent to the energy stored in approximately $30\,m^3$ of natural gas. By lightning, the electric potential in the cloud and on the ground are temporarily equalized. From history, we may think of Benjamin Franklin (1706–1790) and his famous kite-in-a-thunderstorm experiment, one dark afternoon in 1752 (although we are not sure he actually did it). There were others, such as Thomas-Francois Dalibard (1709–1778) near Paris, who around the same time made observations to establish the electric nature of lightning. Less sensational but still startling electric shocks can sometimes occur when we comb our hair or pull off a woollen sweater. The fact that friction causes charging reveals the electrical structure of matter. The building blocks of matter, molecules and atoms, contain electrically charged elementary particles.

The electrical state of an object is characterized by its charge, positive or negative. If we have a free electric charge in the vicinity of a charged object, it will move according to the Coulomb force, proportional to the product of the charges and inversely proportional to the square of the distance between the two charges. The direction of the force lies along the connecting line between the two charges. Therefore, it is very similar to the gravitational force, except that the Coulomb force may be either repulsive or attractive.

Second Law: Magnets

If a sheet of paper is laid on top of a bar magnet and we sprinkle iron filings on it, we will see these tiny needles aligning themselves so that their long axes are parallel, producing a visualization of "the lines of force," which for a magnet go from the south pole to the north pole. Such magnetic field lines are always closed, beginning and ending at the poles of the magnet. Like the force between electric charges, the force between magnets is quite similar to gravity.

The main difference with electric charges is that a magnet always has a south pole if it has a north pole. To the best of our knowledge, there are no separate magnetic charges (or monopoles), but always linked north and south poles. We can break a magnetised rod into two parts and a north and south pole will be automatically created on each part.

Some materials are naturally magnetic. Iron, cobalt, and nickel have the property of being able to attract or repel each other according to their respective positions. Cork or paper are not magnetic. The Earth itself is a giant magnet (as would eventually be understood much better in terms of the dynamo effect proposed by the same Walter Elsasser to be encountered in Chap. 6 on X-ray diffraction). We see this in the motions of a compass needle.[1] When a magnet is freely suspended, its north pole will point toward the Earth's south magnetic pole, which is located near the geographic north pole, whence this south magnetic pole is actually close to the North Pole of the Earth.

---

[1] In 1884, Albert Einstein received a compass from his father. It is said that the young Albert could not stop puzzling over the nature of what he saw.

Dynamical Laws

The next two laws add dynamic aspects to the laws of electricity and magnetism. In doing so, they demonstrate how electrical and magnetic phenomena are actually connected. That is the real subject of the present chapter.

The Third Law is Faraday's law of induction: a time-varying magnetic field creates an electric field. Or, the work per unit charge needed to move a charge around a closed loop is equal to the rate of change of the magnetic flux through any enclosing surface. We can use this to generate electricity. A rotating bar magnet creates a changing magnetic field, which in turn generates an electric field in a nearby wire.[2]

The Fourth Law is Ampére's Law with Maxwell's Addition. An electric current or changing electric fields create a magnetic field. That an electric current generates a magnetic force was discovered in 1819 by Hans Christian Oersted (1777–1851) Oersted (1820). When a compass was set next to a wire carrying a current, its needle turned to a position perpendicular to the wire. This magnetic force differed significantly from what had been seen before: it was not just a force along the straight line between the objects, as for Newton's gravity or Coulomb's law.

Oersted's discovery was the first connection to be found between electricity and magnetism. The first demonstration of a mechanical force produced by the interaction of an electric current and a magnetic field was used to generate a rotary motion by Faraday in 1821. Joseph Henry (1797–1878) was the champion in making strong electromagnets, i.e., magnets in which the magnetic field is produced by an electric current. In doing so, he followed the example of William Sturgeon (1783–1850), known as the "Electrician," who reported the first artificial electromagnet in 1824 in a paper presented to the British Royal Society of Arts, Manufactures, and Commerce.

> The beauty of electricity or of any other force is not that the power is mysterious, and unexpected, touching every sense at unawares in turn, but that it is under law, and that the taught intellect can even now govern it largely.
> — M. Faraday

3.2 Induction

Chance discoveries play a major role in all branches of science. In chemistry, William Perkin (1838–1907) accidentally concocted the first synthetic dye in 1856 when attempting to create, of all things, a cure for malaria. In medicine, a good example is Alexander Fleming's (1881–1955) serendipitous discovery of penicillin in 1928.

---

[2] The electric guitar was invented in the early 1930s by George Beauchamp (1899–1941). Here we need some *string* theory. The point is that the metal strings of the guitar generate an electric current when they move, in the same way as a light can be powered by a dynamo. The strings are partially magnetized by the presence of a larger magnet. When they vibrate, a very small electric current flows through pickup coils. The pickups connect with an electrical circuit and amplifier, which boosts the small electric current and sends it on to a loudspeaker.

And in engineering, the idea of a microwave oven came to Percy Spencer (1894–1970) after noticing that the active radar he was working with, melted the chocolate candy bar in his pocket.

Physics has had its own share of discoveries where an opportunity suddenly arose for genius to act upon, and one of those events set off the unification of what appeared until then to be independent phenomena, namely electricity and magnetism. In 1820, while presenting a demonstration of some electrical phenomena, Oersted noticed that, just as he connected up a galvanic circuit with a wire, the needle of a nearby compass turned away from its stable position pointing north. Upon further investigation, Oersted formulated his eponymous law, stating that an electric current creates a magnetic field around it. That was the first connection found between electricity and magnetism, and the first law linking the two. The law of induction, describing a relation in the other direction, would follow a decade later.

Inspired by Oersted's discovery, Michael Faraday set out to conduct his own investigations in electromagnetism. Although he lacked a formal education in mathematics, Faraday was above all an excellent experimentalist, and on top of that a great theoretician. The description of Faraday as a theorist may be surprising. However, it is inspired by the words of James Clerk Maxwell when studying Faraday's experimental work and his invention of lines of force:

> As I proceeded with the study of Faraday, I perceived that his method of conceiving the phenomenon was also a mathematical one, though not exhibited in the conventional form of symbols. Faraday is a mathematician of a very high order.
> —J.C. Maxwell (1873)

For Maxwell, apparently, the symbols and the formal manipulations are not at all the most important part of mathematical physics. What actually constitutes the soul or programmatic aim of mathematical physics often gets blurred by formal appearances, and then mathematical physics can soon degrade into some sort of theoretical mathematics.

Faraday's motivation becomes clear when we look at his notebook *Chemical Notes, Hints, Suggestions and Objects of Pursuit* (1822), in which he explores possibilities and questions of interest. The chapter on electricity starts with the simple phrase "Convert magnetism into electricity." We already noticed how, in 1821, Faraday made an electric motor on the basis of Oersted's discovery.

In the summer of 1831, the law of induction was discovered. Faraday wrapped two coils of wire around opposite sides of an iron "torus" (a thick ring), taking care to insulate them from each other and from the torus. One coil was plugged into a galvanometer (a device for measuring electric current, fittingly based on Oersted's laws), and Faraday connected the other coil to a battery. A transient current was created, like a "wave of electricity," as Faraday would call it, whenever he connected the wire to the battery, and another when he disconnected it. This was induction, due to the change in magnetic flux that occurred when the battery was connected and disconnected.

Faraday discovered several other realizations of induction in the following months. This was crucial for the development of the concept of electromagnetic field. For example, transient currents were also observed when he slid a bar magnet in and out of a coil of wires. As an ingenious application Faraday managed to generate a steady current by rotating a copper disk near the bar magnet, using a sliding electrical lead.

Faraday explained induction using the concept he called lines of force, which in the hands of Maxwell became the concept of the electromagnetic field.

3.3 The Field Concept

In Newtonian classical mechanics, the main categories are objects (bodies, particles, etc.) and forces (interactions). Newton only considered contact forces, mathematical abstractions of our intuitions about collisions between objects. That gravity seemed to operate instantaneously over immense distances was a significant thorn in Newton's side, and this question would take centuries to be resolved. In electromagnetism, the solution appears as a new category, namely the field concept:

> It is a program that may appropriately be called that of Maxwell: the description of the Physical Reality by means of fields which, without singularities, satisfy a collection of partial differential equations.
> — A. Einstein, At the centenary of the birth of Maxwell, in 1931.

The idea of a field was already known as a description of the values of a physical variable over space and time. For example, we can map out the temperature across Belgium at any given time, and in a suitable limit, about every square centimeter would get a number (the local temperature T), making a field $T(r)$, where r is a spatial variable (the location in Belgium). This field would be continuous for all practical purposes, and in fact it would satisfy differential equations such as the heat equation, together with initial and boundary conditions: *Et ignem regunt numeri.*[3] Similarly, we can speak about the velocity field, say in a river, associating an arrow with an appropriate length to each place in the water. Such fields have a carrier (an object) and are in the first place mathematical tools for description, like statistical tables or for bookkeeping purposes.

The fields in electromagnetism (electric and magnetic fields) differ in at least two important respects. First, they need not be carried by or associated with objects ; they still make sense in empty space, i.e., they can exist independently of a specific medium. Secondly, they have a different status; they are real. The physics is entirely in the fields.

We can replace the objects (charges, wires, magnets, etc.) and have the same effects as long as the electromagnetic fields are the same. They represent the action and reaction, and replace the idea of forces between objects. If a coil and another— ordinary—magnet have the same field, all their electrical and magnetic actions and influences will be identical. Additionally, the interaction between two coils (attraction or repulsion) happens just like with magnets. Therefore, we conclude that the

---

[3] "And even fire is governed by numbers," attributed to Plato and written by Joseph Fourier on the title page of his *Théorie Analytique de la Chaleur* (1822).

field has all the properties of the material carriers. We can even ignore the carriers completely and work solely with the fields. The field itself turns out to be essential; the differences between the sources are insignificant. This unification and reduction of the phenomenology gives the field concept a certain reality.

> Before Maxwell people conceived of physical reality — in so far as it is supposed to represent events in nature – as material points, whose changes consist exclusively of motions, which are subject to total differential equations. After Maxwell they conceived physical reality as represented by continuous fields, not mechanically explicable, which are subject to partial differential equations. This change in the conception of reality is the most profound and fruitful one that has come to physics since Newton; but it has at the same time to be admitted that the program has by no means been completely carried out yet.
> — Albert Einstein (1931), in a Commemoration volume for Maxwell, "Maxwell's Influence on the Evolution of the Idea of Physical Reality."

We also get a methodological advantage. The field concept sets us on the trail of new laws and generalizations. An electric current creates a magnetic field. There are no magnetic currents, but we can now have a changing magnetic field, and we can infer that it will generate an electric field. That is indeed Faraday's law of induction (Maxwell's third law): a changing magnetic field generates an electric field. We can use that principle expressed in terms of fields to generate electricity in many different "material" ways. For example, let us take a coil that is a closed loop and bring a magnet nearby. If we change the position of the magnet, for example, moving it in or out of the coil, we anticipate that an electric current will be generated in the coil for a short time. Such a current means that there must be an electric field that moves the charge carriers. Exactly the same effect will be obtained when a current-carrying coil is set up in the vicinity of a closed coil. Again, it is enough to think in terms of the fields, not the sources.

Finally, by analogy with the foregoing, Maxwell's fourth law comes out naturally: not just electric current but also a change in the electric field will cause a magnetic field. That was Maxwell's specific theoretical prediction. Indeed, that insight is the cornerstone of Maxwell's laws and is an essential step toward predicting the existence of electromagnetic waves:

> The theory I propose may therefore be called a theory of the *Electromagnetic field*, because it has to do with the space in the neighborhood of the electric and magnetic bodies, and it may be called a *Dynamical* Theory, because it assumes that in that space there is matter in motion , by which the observed electromagnetic phenomena are produced."
> — J.C. Maxwell, beginning of the paper "A dynamical theory of the electromagnetic field" (1865).

Let us reiterate the main step. The nature of electromagnetic phenomena meant a departure from the idea that the universe is a spacetime with "discrete stuff" in it. At least, it became much more useful and easy to think of realities differing from particles or "atoms" localized in space. Instead, the work of Faraday and Maxwell gave evidence of the real existence of delocalized physical fields, i.e., spread throughout space and time. As so many things, they are not directly visible to us, but only through their effects, for example by using a "test charge." Test charges can be electrons for the electromagnetic field and tiny masses for the gravitational field.

Note that there is an interesting catch that follows and that has influenced much of the development of physics: the test particles create their own field, which again influences the test particle. Those self-energies and feedback mechanisms require a careful and sometimes subtle calculation to avoid singularities. That is not unrelated to our discussion in Chap. 9 on the frenetic vacuum.

A second and related catch is that fields require quantization, which goes back in the direction of "particle-like" properties. The challenge is to obtain a consistent theory in which one has matter and charges living happily together with the fields. The number of degrees of freedom can be wildly different when the fields are not quantized, giving rise to incorrect entropies and a misunderstanding of the statistical mechanical equilibrium properties. Such (ultraviolet) catastrophe helped bringing about the birth of quantum mechanics.

3.4 Electromagnetic Waves

If, roughly speaking, electricity can induce magnetism and magnetism can be converted into electricity, it is natural to consider the possibility of this process going on forever, with the two fields leap-frogging one another, so to speak. With the full mathematical machinery of his new theory, Maxwell was able to work out this concept of an electromagnetic wave in 1862. When he plugged in the numbers to calculate the velocity with which these joint disturbances in the field would spread out, he was in for a surprise:

> The velocity of transverse undulations in our hypothetical medium, calculated from the electro-magnetic experiments of MM. Kohlrausch and Weber, agrees so exactly with the velocity of light calculated from the optical experiments of M. Fizeau, that we can scarcely avoid the inference that *light consists in the transverse undulations of the same medium which is the cause of electric and magnetic phenomena*
> — James Clerk Maxwell, in *On Physical Lines of Force*, part II, pg. 22

The existence of these waves was proven by Heinrich Rudolf Hertz (1857–1894) about a decade later. In 1886, as another fine example of a serendipitous discovery, Hertz noticed when experimenting with coils that the discharge of a Leyden jar (a primitive capacitor) in one could produce a spark in the other. With this in mind, he conceived of an experiment in which an induction coil delivered high-voltage spikes to a pair of copper wires, serving as a dipole antenna. His receiver was a simple loop of wire with a micrometer spark gap. Hertz was able to verify the finite speed of propagation of these waves by reflecting them off a metal sheet. This allowed him to determine the wavelength of the oscillations, and knowing the frequency, he could then reconstruct the wave velocity, finding once again a close match with the speed of light.

These theoretical and experimental discoveries can also be seen as crucial steps in the unification of electric, magnetic, and optical phenomena. For the electromagnetic field, there was the additional problem of how to think about its medium, the so-called æther. This was crucial in the development of special relativity and of spacetime theories more generally. With the understanding of electromagnetic waves, the problem became acute. The Michelson–Morley experiment, to be discussed in Sect. 8.1 was a wonderfully unsuccessful attempt to discover the motion of the Earth relative to the "light medium." It implies that the same laws of electrodynamics and optics will be valid for all frames of reference for which the equations of mechanics hold good.

3.5   Unification

Physics is not a collection of loosely connected stories. There is a clear ambition to unify things, be it through general statistical laws, or though the recognition of a few basic laws and fundamental constituents. In this way, we can speak about the "essence of Nature," in the sense of its ontology, which gets recognized through such a unification. What physics ends up calling "real," regarding the basic entities and dynamics, is determined by that unification, and from this emerge the many and varied manifestations we encounter in everyday phenomena and experience.

Newton unified the sublunar with superlunar physics, giving us a *universal* theory of gravitation. William Rowan Hamilton (1805–1865) unified classical mechanics with geometrical optics,[4] crucially inspiring for Schrödinger's derivation of the equation that bears his name. Faraday and Maxwell were responsible for the unification of our electric, magnetic, and optical experiences. However, some good glue was still missing, and that was eventually provided by Einstein's 1905 paper on the principle of relativity. Let us listen to its wonderful overture:

> It is known that Maxwell's electrodynamics — as usually understood at the present time — when applied to moving bodies, leads to asymmetries which do not appear to be inherent in the phenomena. Take, for example, the reciprocal electrodynamic action of a magnet and a conductor. The observable phenomenon here depends only on the relative motion of the conductor and the magnet, whereas the customary view draws a sharp distinction between the two cases in which either the one or the other of these bodies is in motion. For if the magnet is in motion and the conductor at rest, there arises in the neighborhood of the magnet an electric field with a certain definite energy, producing a current at the places where parts of the conductor are situated. But if the magnet is stationary and the conductor in motion, no electric field arises in the neighborhood of the magnet. In the conductor, however, we find an electromotive force, to which in itself there is no corresponding energy, but which gives rise — assuming equality of relative motion in the two cases discussed — to electric currents of the same path and intensity as those produced by the electric forces in the former case.
> — Albert Einstein, From the introduction of "On the electrodynamics of moving bodies" (1905). English translation of his original *Zur Elektrodynamik bewegter Körper*, in Annalen der Physik **17**, 891 (1905) which appeared in the book *The Principle of Relativity*, published in 1923 by Methuen and Company, Ltd. of London.

---

[4] Note that the letter H used to denote the Hamiltonian stands for "Huygens", as used by Lagrange in his book *Mécanique Analytique*, written before Hamilton worked on mechanics. In another way, Hamilton was also a "unifier" of poetry and science, as can already be seen in his poem "The Enthusiast" of 1826.

Einstein gives examples to show how phenomena of electrodynamics are not compatible with Newton's ideas of absolute rest. With this "principle of relativity," a much further unification of electromagnetism is achieved.

Chapter 4
Looking at Fluctuations

Discreteness has pleasant consequences. If something is or can be in (only) a finite number of configurations, we can truly count possibilities. If the numbers get large, as in macroscopic collections of particles, we can understand what physically coarse-grained state would be most typical, by hypothetically using some version of Laplace's indifference principle.[1] That is the origin of the notion of entropy, as understood by Ludwig Boltzmann (1844–1906). In addition, there may be deviations from what is typical. These are fluctuations, and we can see them, and feel them.

4.1 Natural Philosophy

Newton's great work, the *Principia*, explains the mathematical foundations of natural philosophy. The latter has a long tradition, and its objective is to discover the constitution and laws of the physical world. Finding out what exists in the physical world is of course a matter of combining empirical research with scientific theory, but natural philosophy is mostly concerned about the fundamental implications of that combination.

Natural philosophy then is a philosophical reflection on Nature, and more specifically, it is the study of foundational aspects of the natural sciences. It has mostly been developed to study the philosophy of physics, but its ambition today also stretches to aspects of biology (evolution), chemistry (life), computer science (information), and other areas.

For Europe, the intellectual tradition started in Ancient Greece in the Milesian school founded in the sixth century BC, flourishing in the pre-Socratic philosophy

---

[1] Given constraints and available evidence, we should distribute our "degree of belief" equally among all the remaining possible outcomes.

© The Author(s), under exclusive license to Springer Nature Switzerland AG 2023
C. Maes, *Facts of Matter and Light*,
https://doi.org/10.1007/978-3-031-33334-7_4

and then, in the Hellenistic period, with Archimedes as its culminating figure and the father of mathematical physics.

The rather rapid birth and development of a scientific culture in Ancient Greece in the sixth to fourth centuries BC was steered by a new ambition of explaining Nature through Nature.[2] That intellectual project was perhaps still largely speculative and theoretical, but differed essentially from Babylonian or Ancient Egyptian efforts, which typically dealt only with collecting data, then fitting simple models, often directly motivated by economic interests or by the preservation or protection of those in power. Somewhat like what is sometimes claimed to be part of the scientific output today. Yet, good science rocks, i.e., always has something revolutionary about it. In that way, the Greek natural philosophers reached much further. Here we had pre-scientists who were not in the service of kings or money, discussing nature devoid of supernatural causes or of mythical origins, representing the substance and workings of Nature even in terms of unobservable reality, but in terms of mechanisms and causes that referred at least in principle to observable quantities.

Here, the second meaning of the word "nature" pops up, referring to basic or inherent features. It is the ambition of talking foundations to aim toward the *true* nature of Nature. We find conjectures about the ultimate structure of all matter (and mind) with serious attempts to reduce and unify under slogans like *All is water* (Thales) or *All is air* (Anaximenes). This marks the beginning of an effective program to connect the multitude of known phenomena through a small number of simple principles, ingredients, and mechanisms. Condensation and dilution were, for example, assumed to transform between different macroscopic phases of matter. We shall return to this line of thinking in Chap. 10. For what matters here, literally, is the corpuscular nature of matter.

4.2 Atoms

It was in this cultural climate, referred to as the Greek Miracle, that the idea of atoms first took off.[3] More specifically, in the Western world, the idea of atoms goes back to Ancient Greece, with philosophers like Leucippus,[4] Democritus, and Epicurus.[5] We are here in the period 450–400 BC.

---

[2] In his "Nature and the Greeks," Erwin Schrödinger remarks that *doing science is thinking and speaking like the Greeks.*

[3] At least in Western science. There are many other sources and developments of natural philosophy. For example, Indian atomism was discussed by the Nyaya-Vaisheshika philosophers, and their roots can be traced back to the Upanishadic doctrine of five elements (pancha-mahabhutas). Other kinds of atomism are found in the Buddhist and Jaina worlds views.

[4] Aristotle and Theophrastus are the sources for explicitly crediting Leucippus with the invention of atomism. On the other hand, Epicurus comments that there was no such philosopher Leucippus
...

[5] Best known from the influential poem "De Rerum Natura," written by Lucretius.

Nothing exists except atoms and empty space, everything else is (just) speculation.
—Democritus

These were the first in Europe to attempt a program to explain everything through the mechanism of attraction and repulsion of small particles. Its far-reaching consequences for the understanding of the world are captured in the following much more recent and more dramatic suggestion:

> If, in some cataclysm, all of scientific knowledge were to be destroyed, and only one sentence passed on to the next generations of creatures, what statement would contain the most information in the fewest words? I believe it is the atomic hypothesis (or the atomic fact, or whatever you wish to call it) that all things are made of atoms, little particles that move around in perpetual motion , attracting each other when they are a little distance apart, but repelling upon being squeezed into one another. In that one sentence, you will see, there is an enormous amount of information about the world, if just a little imagination and thinking are applied [...]
> — Richard Feynman in *Lectures on Physics, 1961–1962.*

The views of the atomists were largely ignored for a very long time. They were already characterized by Aristotle as mere materialists because they stated that the mind was also made up of atoms. Atomic theory only re-emerged in our region of the world around the time of the Renaissance. Special mention should go to Isaac Beeckman (1588–1637), Sébastien Basson (ca. 1573–1630), and Pierre Gassendi (1592–1655) who contributed to the reintroduction of atomism. They helped to introduce the idea of atoms into modern science and make the atomistic conception compatible with the Western Christian teaching.

From that time on, the atomic hypothesis provided an important working framework for the birth of chemistry. John Dalton (1766–1844), Joseph Gay-Lussac (1778–1850) and Amedeo Avogadro (1776–1856) opened the way to modern chemistry. Even well into the nineteenth century, it was mainly chemists who thought about atoms in purely practical terms in order to visualize molecules.[6]

Take for example Jacobus H. van 't Hoff (1852–1911), who used models of atoms and molecules to understand the properties of a substance via its (discrete) spatial structure. It was also van 't Hoff who noted the analogy between the molecules or granules in a solution and gas molecules. The molecules of a solution form a kind of "gas" for which the ideal gas law yields a formula for the osmotic pressure. Einstein discusses precisely this case of osmosis in the first paragraph of his 1905 article on Brownian motion, where he demonstrated the existence of atoms "by counting them."

How indeed did we eventually convince ourselves about the reality of atoms and molecules? Much was needed to get there, and there was also multiple evidence, but here we focus on the work and experiment of Perrin. Jean Baptiste Perrin (1870–1942) did the experimental work to test and verify Einstein's predictions concerning Brownian motion, thereby settling the century-long dispute. We come back to it below. Perrin received the Nobel award in 1926 for that and other work *on*

---

[6] Typically, Jean-Baptiste Biot (1774–1862) and Louis Pasteur (1822–1895) are mentioned as the first stereochemists.

the discontinuous structure of matter, which put a definite end to the long struggle regarding the question of the physical reality of molecules.

Looking at it from a bigger distance, the essential ingredient was the study of fluctuations. When Brownian motion was understood and quantitatively characterized as a dynamic fluctuation phenomenon, atoms and molecules could be counted and hence became part of our physical world.

Since then, fluctuations have gained importance, being both visible and revealing. Today, for example, our best telescopes can track fluctuations in the energy density of the early universe and the cosmic background radiation. At a totally different scale, single-particle tracking techniques can explore biomolecular dynamics in living cells with single-molecule sensitivity and nanometer spatial resolution. For vacuum polarization, particle-antiparticle pairs are spontaneously created giving rise to a fluctuating vacuum (see more in Chap. 9). Also the idea of a virtual particle in condensed matter theory often refers to a fluctuation.

4.3 Limit Theorems

The discrete cannot be easily reconciled with the continuous. This tension between continuous and discrete lives on in various other relations as well, such as between the real and its model, the real and the ideal, and between physics and mathematics. Many paradoxes attest to the conceptual struggle that accompanied the development of the limit concept, the infinite, and the transition between discrete and continuous descriptions.[7]

Such questions also arose when trying to reconcile the corpuscular nature of matter with continuous motion. The invention of analysis (integration and differentiation) was the important step for classical mechanics. A related issue was the reconciliation of atomic theory with a theory in which the properties of space vary continuously. This question arose regarding the production and transformation of light, the theme of Einstein's first publication[8] from 1905, where it is explicitly mentioned in the first paragraph.

The idea of atoms (or more generally molecules, particles, or corpuscles) immediately gives rise to the idea of fluctuations. The impacts of atoms will fluctuate because each atom will have its own motion and collisions on the microscopic level. They are moving all the time due to their kinetic energy, and this motion will impact

---

[7] Related discussions were very much alive among the Ancient Greeks. See, e.g., *Nature and the Greeks* by Schrödinger, Cambridge University Press (1954), for a less well-known paradox described by Democritus: Suppose a cone is cut in half by a plane parallel to its base. Now we have two pieces, a smaller cone and a cone without a top, which when put together restore the original cone. Now look at the two circular cross-sections that previously touched — are their areas equal or not? How could they be different, and yet how could they be equal?

[8] "A heuristic point of view of the production and transformation of light" was revolutionary because it reintroduced the idea that light is composed of particles, later called photons.

on any kind of probe immersed among them. The result is a fluctuating (stochastic) motion of the probe, provided that it is small enough, e.g., a mesoscopic probe.

> In the case of a surface having a certain area, the molecular collisions of the liquid which cause the pressure, would not produce any perturbation of the suspended particles, because these, as a whole, urge the particles equally in all directions. But if the surface is of area less than is necessary to ensure the compensation of irregularities, there is no longer any ground for considering the mean pressure; the inequal pressures, continually varying from place to place, must be recognized, as the law of large numbers no longer leads to uniformity; and the resultant will not now be zero but will change continually in intensity and direction. Further, the inequalities will become more and more apparent the smaller the body is supposed to be, and in consequence the oscillations will at the same time become more and more brisk...
> J.B. Perrin, (Perrin, 1910), page 4.

Three levels of description are on stage, the microscopic one which is the scale of fluid molecules, the mesoscopic scale of the colloidal particle or probe, and the macroscopic description of diffusion. The question is then to reconcile them, the more microscopic description of particles with the regularity, macroscopic determinism, and continuity of our everyday experiences. The important insight here is that these indeed different levels of description, a hierarchy so to speak, gives rise to seemingly independent layers of autonomy. For understanding that autonomy, one crucial step is simply to consider the sheer number of atoms. The main example in physics, the transition to thermodynamics (and its corrections), could now be formulated as similar to what happens in the limit laws of probability theory.

To begin with, here is (an example of) the law of large numbers:

> Consider N independent and identically distributed random variables X_i, $i = 1, \ldots, N$, where, for simplicity, each has three possible values $X_i \in \{-1, 0, 2\}$, and $\text{Prob}[X_i = -1] = p_{-1}$, $\text{Prob}[X_i = 0] = p_0$, and $\text{Prob}[X_i = 2] = p_2$. Then, with probability tending to one as $N \uparrow \infty$, the sums converge
>
> $$\frac{1}{N}(X_1 + X_2 + \cdots + X_N) \longrightarrow \langle X \rangle = -p_{-1} + 2p_2, \qquad (4.1)$$
>
> and the deterministic limit is the statistical average $\langle X \rangle$.

What is large in a bottle of gas, a glass of water, or a brick is the number of particles making it up. They are counted in multiples of Avogadro's number, denoted by $N = N_A = 6.022 \times 10^{23}$, which is the number of units per mole of a given substance. In a bottle of air, there will be about $N = 10^{22}$ molecules.

Fluctuations can be viewed as corrections to that law of large numbers. They give the deviations from, or the noise, around the average. Consider again the sum $X_1 + X_2 + \cdots + X_N$; it is about equal to $N \langle X \rangle + O(\sqrt{N})$, and we are interested in characterizing that correction of order $\sqrt{N}$.

For example, we can blow up by $\sqrt{N}$ the difference in (4.1), obtaining

$$\xi_N = \sqrt{N} \left[\frac{1}{N}(X_1 + X_2 + \cdots + X_N) - \langle X \rangle \right].$$

The distribution of ξ_N in the limit $N \uparrow \infty$ tends to a zero mean Gaussian variable with variance equal to the variance of X: $\sigma^2 = \text{Var}[X] = (p_{-1} + 4p_2) - (-p_{-1} + 2p_2)^2$.

For no matter what numbers $a > b$, as $N \uparrow \infty$,

$$\text{Prob}\big[\xi_N \in [a, b]\big] \to \int_a^b \mathrm{d}x \, \frac{1}{2\pi\sigma} e^{-x^2/2\sigma^2}. \tag{4.2}$$

which is a convergence of the probability distribution. This is called the central limit theorem, a truly universal result, sometimes called the law of error.[9]

One can also consider large (or larger) deviations, and ask for the probability

$$\text{Prob}\left[\frac{1}{N}(X_1 + X_2 + \cdots + X_N) \approx a\right] \sim e^{-NI(a)}, \tag{4.3}$$

for a given value a[10] The idea is that, when $a \neq \langle X \rangle$, there is a large deviation from the law of large numbers (4.1), and the probability gets exponentially small as $N \uparrow \infty$. The so-called large deviation function $I(a) \geq 0$, and is zero only when $a = \langle X \rangle$: $I(\langle X \rangle) = 0$.

There appears something constructive as well, upon reversing the above view: if we know the function I in (4.3), we can actually find $\langle X \rangle$ by minimizing $I(a)$ over all possible a, i.e., the minimizer is the typical value $\langle X \rangle$. We now get nothing less than a method for finding out what to expect: what is *typical* can be found by minimizing the large deviation function I. This is the origin of variational principles in thermodynamics, used to understand what behavior we should expect. The mathematics is slightly more complicated but the logic remains unaltered. Will that piece of matter turn out to be a piece of graphite or will it form a diamond at specific values of temperature and pressure? It is the large deviation rate function I that decides. It is called the Gibbs free energy in thermodynamics.

> The true logic of the world is the calculus of Probabilities, which takes account of the magnitude of the probability which is, or ought to be, in a reasonable man's mind.
> — J.C. Maxwell.

At the end of the nineteenth century, in the hands of physicists like James Clerk Maxwell and Ludwig Boltzmann, and inspired by statistical considerations,[11] the

---

[9] Large numbers do play their role in the Comedy of Errors, where it is asserted that "I to the world am like a drop of water That in the Ocean seeks another drop." (Act 1, Scene 2).

[10] The notation $\simeq a$ is not explained here, but we can think of the normalized sum $(X_1 + X_2 + \cdots + X_N)/N$ to sit in a small interval around the number a. The meaning of $\text{Prob} \sim \exp(-NI(a))$ is to take logarithms left and right, after which we divide by N.

[11] Adolphe Quetelet (1796–1874) had an important role here, introducing statistics into sociology, as we know it still today. His main essay, *Physique sociale*, describes the concept of the "average man" and the unifying role of the normal distribution. Maxwell would also use analogies between social studies and the kinetics of gases. As a matter of fact, Charles Darwin, Francis Galton, and James Clerk Maxwell were all aware of Quetelet's work, and aware also that they had something to learn there. For example, Maxwell read a commentary on Quetelet's achievements in an essay by John Herschel in the Edinburgh Review. After all, according to statistical mechanics, thermodynamics can be viewed as an instance of discovering "population phenomena," i.e., large-scale regularities that transcend microscopic or individual differences.

combination of mechanics and atomism formed the basis of kinetic gas theory and statistical mechanics. Statistical mechanics was born here, as the title of the book of 1902 by Josiah Gibbs (1839–1903), having far-reaching consequences for the development of a wide range of scientific disciplines, including chemistry of course, and for the language of theoretical physics as spoken e.g. in quantum field theory.

> It is one of the striking features of the work of Gibbs, noticed by every student of thermodynamics and statistical mechanics, that his formulations of physical concepts were so felicitously chosen that they have survived 100 years of turbulent development in theoretical physics and mathematics
> — A.S. Wightman, 1990

From the above limit theorems and from the large deviation theory as exemplified in (4.3), statistical interpretations became possible and deviations from thermodynamics could be quantified as well. There, the concept of entropy, first introduced by Rudolf Clausius (1822–1888) as the state function whose differences give the reversible heat over absolute temperature, got a statistical meaning, and the second law of thermodynamics got the status of a statistical law. That was mainly achieved by Ludwig Boltzmann, (Bricmont, 2022). An early example was the consideration of so-called Maxwell demons, especially for systems consisting of not so many particles,[12] that keep a system from not reaching its typical behavior. Historically, however, the main example of a correction to thermodynamics, namely the correction (4.2) to the law of large numbers, served as proof for the existence of atoms, to which we turn next.

4.4 Brownian Motion

There is a long and well-documented history of Brownian motion, (Nelson, 1967). While observed by various people before (such as in 1785 by Jan Ingenhousz (1730–1799), and earlier by Antoni van Leeuwenhoek (1632–1723)), the name refers to the systematic studies of Robert Brown (1773–1858), who observed the endless, irregular, unpredictable, and universal motion of micrometer-sized particles suspended in fluids at rest. This motion becomes more erratic as the particle size decreases, the temperature increases, or the viscosity decreases. Brown theorizes that matter is

---

[12] The "finite being" that James Clerk Maxwell introduced in a letter to his good friend and colleague Peter Guthrie Tait (1867) has become known as Maxwell's demon. Its purpose was to "pick a hole" in the second law of thermodynamics; it was not so much to raise the possibility that ingenuity (of supernatural beings like demons or angels) could violate that thermodynamic law, but to point out aspects of its conceptual status. As is well known, spontaneous fluctuations can cause energy or particles to flow in the "wrong" direction, and this is less improbable for a smaller number of particles and over shorter times. By "wrong," we mean here "in apparent contradiction" with the second law. It was Lord Kelvin (William Thomson) (1824–1907) who first used the name "demon" in 1874. It appears that Maxwell himself did not like the name. To Tait he declared "Call him no more a demon but a valve." He also explained in this note "Concerning Demons" that his purpose was "to show that the second law of thermodynamics has only a statistical certainty."

composed of *active molecules*, exhibiting a rapid, irregular motion having its origin in the particles themselves and not in the fluid in which they are suspended. Interestingly, about 150 years later, people started to speak about *active matter*, e.g., to describe the locomotion of bacteria such as *E. Coli* or of optically driven colloids, and it is quite clear that the motion of a probe surrounded by active particles differs significantly from that of Brownian motion, Wu and Libchaber (2000); Maes (2020).

There is of course no *vitalist* origin[13]; the motion is caused by the collisions of the tiny particles with the fluid particles. What we observe on that mesoscopic scale is a reduced dynamics, having no direct access to the actual mechanics of the fluid particles. In that sense, the theory of Brownian motion provides a kind of microscope for visualizing molecular motion. It was thus a convincing and important ingredient in leading us to an understanding of the corpuscular nature of matter, i.e., the atomic hypothesis:

> I am now convinced that we have recently become possessed of experimental evidence of the discrete or grained nature of matter, which the atomic hypothesis sought in vain for hundreds and thousands of years. The isolation and counting of gaseous ions, on the one hand, which have crowned with success the long and brilliant researches of J.J. Thomson, and, on the other, agreement of the Brownian movement with the requirements of the kinetic hypothesis, established by many investigators and most conclusively by J. Perrin, justify the most cautious scientist in now speaking of the experimental proof of the atomic nature of matter.
> — Wilhelm Ostwald, Grundriss der allgemeinen Chemie (4th edn., 1909)

The understanding of Brownian motion as a fluctuation-induced phenomenon, thus also correcting and extending thermodynamics, is not at all restricted to motion in fluids. For example, over the years 1938–1943, the work of Subramanyan Chandrasekhar (1910–1995) included an important model for dynamical friction in stellar dynamics based on Brownian motion (Chandrasekhar, 1949). Chandrasekhar derived the Fokker–Planck equation[14] for stars and showed that long-range gravitational encounters provide a drag force, dynamical friction, which is important in the evolution of star clusters and the formation of galaxies. The first prediction of a nonzero cosmological constant (in Einsteinian gravity) was made by Rafael Sorkin in the early 1990s, assuming a discrete and fluctuating spacetime structure; see (Sorkin, 2007) for a more recent review.

From an even broader perspective, the study of Brownian motion marks the beginning of stochastic modeling and reveals the obvious role of fluctuations in physics phenomena and beyond. Models of Brownian motion and diffusion processes have been used at least since the work of Louis Bachelier (1870–1946) (PhD thesis *Théorie de la spéculation*, 1900) to evaluate stock options and apply stochastic modeling to the

---

[13] It appears that the first theories of Brownian motion concluded that the particles were alive. Others, perhaps in the tradition of Aristotle, assumed the Brownian motion of pollen witnessed of the active force in sexual reproduction. The first hypothesis of Brown was indeed that the observed motion was not only vital but peculiar to the male sexual cells of plants.

[14] The partial differential equation for the time evolution of probability densities, as first written down by Adriaan Fokker (1914) and Max Planck (1917).

study of finance, culminating perhaps in the famous Black–Scholes–Merton model and Black–Scholes formula (1973) for the price of a European-style option.

4.4.1 The Random Walk as a Diffusion Model

To obtain a particle model for diffusion, we consider a random walk on the integer multiples of some small δ, occupying the sites $0, \pm\delta, \pm2\delta, \ldots$. After each time τ, the walker moves one step to the left or to the right with equal probability. Of course, we bear in mind that we are trying to model the motion of a Brownian particle that is being pushed around at random. Then, the probability of finding the particle at position $n\delta$ at time $k\tau$ equals

$$P(n\delta, k\tau) = \frac{1}{2^k} \frac{k!}{\left(\dfrac{k-n}{2}\right)! \left(\dfrac{k+n}{2}\right)!}.$$

Taking $n = x/\delta$, $k = t/\tau \uparrow \infty$ while fixing $D = \delta^2/2\tau$, we have the continuum limit

$$\lim_{\delta \downarrow 0} \frac{1}{\delta} P(n\delta, k\tau) = \rho(x, t) = \frac{1}{\sqrt{4\pi Dt}} e^{-x^2/4Dt},$$

for the (well-normalized) probability density $\rho(x, t)$, and this does indeed satisfy the diffusion equation

$$\frac{\partial}{\partial t}\rho(x, t) = D \frac{\partial^2}{\partial x^2}\rho(x, t), \tag{4.4}$$

with diffusion constant

$$D = \frac{k_B T}{6\pi\eta d}.$$

Here Boltzmann's constant k_B determines Avogadro's number N_A by the relation $N_A = R/k_B = 6.022 \times 10^{23}$ per mole, via the universal gas constant R. This was indeed the subject of Einstein's thesis in 1905 to determine how many particles of gas there are in 1 mole. Seeing atoms or molecules became equivalent to counting them.

If we consider initial conditions $x_0 \neq 0$ or an initial time $t_0 \neq 0$, we can obtain the fundamental solution of that diffusion equation (4.4) in the form

$$\rho(x, t) = \frac{1}{\sqrt{4\pi D(t - t_0)}} e^{-(x-x_0)^2/4D(t-t_0)}, \qquad \rho(x, t_0) = \delta(x - x_0).$$

We can calculate all moments (taking m even):

$$\langle(x(t) - x_0)^m\rangle = \int_{\mathbb{R}} dx \, (x - x_0)^m \, \frac{1}{\sqrt{4\pi Dt}} e^{-(x-x_0)^2/4Dt} \sim t^{m/2}.$$

This means that while $\langle(x - x_0)^2\rangle$ is of order t, the higher moments $m > 2$ tend to zero faster than t as $t \downarrow 0$.

There is of course another way to get back that solution. The point is that the random walk itself satisfies the master equation

$$P\big(n\delta, (k+1)\tau\big) = \frac{1}{2} P\big((n-1)\delta, k\tau\big) + \frac{1}{2} P\big((n+1)\delta, k\tau\big),$$

or, again with $x = n\delta$, $t = k\tau$, and $D = \delta^2/2\tau$,

$$\frac{P(x, t+\tau) - P(x, t)}{\tau} = \frac{\delta^2}{2\tau} \frac{P(x+\delta, t) + P(x-\delta, t) - 2P(x, t)}{\delta^2},$$

which tends to the diffusion equation (4.4) in the limit $\delta \downarrow 0$ with D held fixed.

Finally, we can look at the process in a more probabilistic way. After all, the position of the walker is the random variable (starting from $X_0 = 0$)

$$X_{k\tau} = \delta\big(v_1 + v_2 + \cdots + v_k\big),$$

where each $v_i = \pm 1$ with equal probability (independent random variables) and $\delta = \sqrt{2\tau D} = \sqrt{2Dt/k}$. We send k to infinity and consider the limit of the sum

$$\xi_k = \sqrt{\frac{2Dt}{k}} (v_1 + v_2 + \cdots + v_k). \tag{4.5}$$

The central limit theorem (4.2) tells us that ξ_k converges in distribution to a Gaussian random variable with mean zero and variance $4Dt$. In fact, more is true. Not only do we get convergence to a Gaussian for fixed time t, but the process itself has a limit and becomes a Gaussian process. That limiting process, the rescaled limit of the standard random walk, is the so-called Wiener process,[15] also simply named after the basic phenomenon it models, Brownian motion.

4.4.2   Sutherland–Einstein Relation

William Sutherland (1859–1911) wrote about Brownian motion and diffusion in a paper which was published in 1904 (Sutherland, 1904), a year before Einstein's paper (Einstein, 1905). However, we shall follow the master himself.

---

[15] Norbert Wiener (1894–1964) pioneered much of the mathematical study of Brownian motion. After World War II, Wiener became increasingly concerned with the militarization of science.

A cloud of spherically shaped Brownian particles of diameter d (or a repeated measurement of the position on a single particle) in a fluid with viscosity η moves according to the diffusion equation (4.3). As in the previous section, we can consider the particle density as $n(x, t) = N\rho(x, t)$ for a total of N particles. The number of particles is conserved and indeed the diffusion equation is a continuity equation:

$$\frac{\partial}{\partial t}n(x, t) + \frac{\partial}{\partial x}j_\mathrm{D}(x, t) = 0, \quad j_\mathrm{D}(x, t) = -D\frac{\partial}{\partial x}n(x, t).$$

The diffusive current j_D is generated by density gradients. Suppose now, however, that there is also a constant external field, such as gravity. We put the x-axis vertically upward. Using Newton's equation, we would write $\dot{x} = v$ and

$$m\frac{\mathrm{d}}{\mathrm{d}t}v(t) = -mg - m\gamma v(t),$$

where $m\gamma$ is the friction (or drag) coefficient and γ is called the damping coefficient. The particle current caused by the external field is $j_\mathrm{g}(x, t) = n(x, t)v(t)$ so, under stationary conditions where $\mathrm{d}v/\mathrm{d}t = 0$, the contribution to the particle current from gravity equals $j_\mathrm{g}(x) = -gn(x)/\gamma$.

However, at equilibrium, there is no net particle current and the diffusive current j_D caused by concentration gradients cancels the "gravity" current. For these currents j_g and j_D, we can use the equilibrium profile $n(x)$. It is given by the Laplace barometric formula

$$n(x) = n(x_0)\,\mathrm{e}^{-V(x-x_0)/k_\mathrm{B}T},$$

where V is the external potential, given by $V(x) = mgx$ for gravity. Hence, we have $j_\mathrm{D}(x) = Dmgn(x)/k_\mathrm{B}T$ and the total current $j = j_\mathrm{g} + j_\mathrm{D}$ vanishes at equilibrium, when

$$j(x) = -\frac{gn(x)}{\gamma} + \frac{Dmgn(x)}{k_\mathrm{B}T} = 0.$$

This implies that the diffusion constant is equal to

$$D = \frac{k_\mathrm{B}T}{m\gamma},$$

which is usually called the Einstein relation (Einstein, 1905). The Stokes relation further specifies $m\gamma = 6\pi\eta R$, giving it in terms of the particle diameter R (at least for spherical particles) and the fluid viscosity η.

Improved mathematical formulations and essential contributions to the systematic understanding of viscosity and its dependence on the density of the solute were important features of the work of George Batchelor (1920–2000) (Batchelor, 1972; Batchelor and Green, 1792), and *An Introduction to Fluid Dynamics* (Cambridge University Press, 1967) is still a classic reference. See also the recent work on Einstein's effective viscosity in (Duerinckx and Gloria, 2023).

Fig. 4.1 Reproduced from
the book *Les Atomes* by Jean
Baptiste Perrin, plots of the
motion of three colloidal
particles of radius 0.53 μm,
as seen under the
microscope. Successive
positions every 30 seconds
are joined by straight line
segments (the mesh size is
3.2 μm)

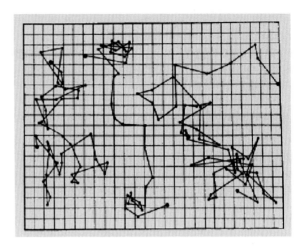

4.5   Perrin's Experiment

We now come to the real thing, that is, the confirmation of everything above; see,
e.g., (Maiocchi, 1990). In 1910, Jean Baptiste Perrin did the experimental work[16]
required to check Einstein's predictions (Perrin, 1910, 1913). He thereby obtained
a consistent and unique number for Avogadro's constant N_A by considering the
Brownian motion of colloidal particles and the analysis above. As it agreed with other
methods for determining N_A, Brownian motion became an important confirmation
of the atomic hypothesis.

Perrin was able to make spherical particles (grains or spherules) of known size
to use as the Brownian particles.[17] He used a state-of-the art centrifuge. The nature
of the liquid (water, solutions of sugar or urea, and glycerol) and its viscosity was
also known. Using the then recently invented ultra-microscope,[18] Perrin watched
the resulting displacements by noting the horizontal projections of the same grain
at the beginning and at the end of a time interval (see Fig. 4.1). He showed that the
density distribution of the particles and their mean square displacement per unit time
and their rotation agreed both with the assumption that the particles were in random
motion and with Newton's laws. The agreement with Einstein's theory was such that

---

[16] Perrin did many other experiments. For example, in 1895 he established that cathode rays (i.e.,
electrons) are negatively charged particles.

[17] The very preparation of the emulsions with particles of a known radius was in fact the most
delicate and complex operation in the whole procedure.

[18] This uses light-scattering, not reflection or absorption, and allows much smaller particles to be
viewed. The colloid is illuminated with an intense light beam ("Tyndall cone"). In 1925, Richard
Adolf Zsigmondy (1865–1929) received the Nobel Prize in Chemistry for his research on colloids
and the ultramicroscope, developed by himself and Henry Siedentopf (1872–1940) while working
for Carl Zeiss AG.

it was impossible to doubt the correctness of the kinetic theory of the translational Brownian movement:

> Thus the molecular theory of the Brownian movement can be regarded as experimentally established, and, at the same time, it becomes very difficult to deny the objective reality of molecules.

The atomic age was about to start, and people soon picked up on the idea of making individual atoms "visible." In fact, the atom was first unveiled in 1911 in the Rutherford experiment (see Sect. 7.2.1), which showed that the atom has a structure.

4.6 Fluctuation–Dissipation Relation

One of the most fascinating and unifying aspects of the physical world is the close relationship between fluctuations (variability and deviations from the most typical behavior) and dissipation (the production of entropy). Heat and friction reduce the availability of energy to perform work (which is dissipation), but this is closely related to fluctuation behavior. In point of fact, the somewhat paradoxical point is that it is *only when* the entropy can decrease now and then through a fluctuation that the entropy can systematically increase at all, e.g., through the relaxation of a macroscopic system to equilibrium. On the other hand, in thermal equilibrium of classical systems, response to an external stimulus can be quantified, e.g., from Kubo or Green-Kubo relations, (Kubo, 1986; Spohn, 1991; Maes, 2020), from the statistical correlation between the induced or excess dissipation and the observable. That is a general consequence of dynamical time-reversibility of the microscopic world combined, at least in the classical case, with the emergence of *memory loss* (also called molecular chaos) which arises when the system is weakly coupled to a huge number of degrees of freedom.

The above can be made quantitative to some extent, and in fact in various ways. One of the oldest and most useful ways usesfluctuation–dissipation relations. The fluctuation–dissipation theorem is itself a cornerstone of linear response theory for equilibrium systems (see (Kubo, 1986; Spohn, 1991)). It generates formulæ that connect a generalized resistance (friction) with the variance of thermodynamic potentials.

> The compartments in which human thought are divided are not so watertight that fundamental progress in one is a matter of indifference to the rest. The great change in theoretical physics which began in the early years of the twentieth century is a purely scientific development, but it must affect the general current of human thought, as at earlier times the Copernican and Newtonian systems have done.
> — A. Eddington

As a general consequence: the internal fluctuations in a system are related to its susceptibility under an applied external field. This is of great practical and theoretical interest. The most immediate example concerns the relationship between diffusion and mobility, with the first studies described above, in the work of Sutherland (1904,

1905) and Einstein (1905). These are early examples of the fluctuation–dissipation theorem.

Another forerunner was the Johnson–Nyquist relation for the voltage fluctuations in electrical impedances. In 1926, while working as an experimental physicist at Bell Labs, John Bertrand Johnson (1887–1970) was researching noise in electronic circuits. He found that, regardless of whether any voltage was applied to a given resistor, there always remained a low level of noise, whose power was proportional to the temperature (Johnson, 1927). The theorist there, Harry Nyquist (1889–1976), developed an elegant explanation similar to what we saw above for the Sutherland–Einstein relation for Brownian motion (Nyquist, 1928). This electronic noise is generated by the thermal agitation of the charge carriers (electrons) inside.

That turns out to be a valid strategy: if we want to know how susceptible an equilibrium system is to a small external stimulus (force or change of parameter), we must measure the correlation between the extra applied potential and the induced entropy flux to a thermal bath. The stimulus can create a (small) current, but never without heat (except in quantum phenomena like superconductivity discussed in Sect. 10.4), and the conductivity (which is a susceptibility) is in fact proportional to the unperturbed correlation between the current and that entropy flux. That is how mobility and diffusion are connected in the Einstein relation, and that is how the resistance is connected with the variance of the voltage in an electric circuit. This kind of fluctuation–dissipation relations, valid as such only close to equilibrium, have a very wide applicability (Callen and Welton, 1951), and have an extension to far-from-equilibrium regimes (Maes, 2020). There, however, one must add a frenetic contribution, i.e., a nondissipative time-symmetric measure of the excess in dynamical activity caused by the stimulus (Maes, 2020).

Chapter 5
Quantization

At the end of the nineteenth century, opinions about the state of physics were rather mixed. On the one hand, there was a widespread feeling that the end of physics was in sight. All problems, provided they received the necessary elaboration, would be reduced over time to questions of thermodynamics, mechanics, or electromagnetism. It was, so to speak, only a matter of time and greater accuracy:

> The more important fundamental laws and facts of physical science have all been discovered, and these are now so firmly established that the possibility of their ever being supplanted in consequence of new discoveries is exceedingly remote [...] our future discoveries must be looked for in the sixth place of decimals.
> — Albert Abraham Michelson (1894, dedication of Ryerson Physical Laboratory, quoted in Annual Register 1896, p. 159)

On the other hand, some scientists were well aware of dark clouds gathering over the physics landscape:

> The beauty and clearness of the dynamical theory, which asserts heat and light to be modes of motion, is at present obscured by two clouds. I. The first came into existence with the undulatory theory of light, and was dealt with by Fresnel and Dr. Thomas Young; it involved the question, how could the Earth move through an elastic solid, such as essentially is the luminiferous æther? II. The second is the Maxwell–Boltzmann doctrine regarding the partition of energy.
> — Lord Kelvin, "Nineteenth-Century Clouds over the Dynamical Theory of Heat and Light." Speech on April 27, 1900 to the Royal Institution. Published in The London, Edinburgh, and Dublin Philosophical Magazine and Journal of Science 2, no. 6, 1–40 (1901).

Between the triumph over the end of physics and the residual internal tensions, it was often felt that electromagnetism and even thermodynamics were more fundamental than mechanics. In thermodynamics, there were the *energeticists* who attempted to reduce all of nature to forms of energy. There was a conjecture that all properties of matter could be derived from its electrical or magnetic composition. There was also the æther hypothesis, that is, the belief in a continuous and pervasive electromagnetic medium. Most people resolutely opted for for the energy, the æther, and the field concepts. This perhaps reflected a more general *Zeitgeist*:

© The Author(s), under exclusive license to Springer Nature Switzerland AG 2023
C. Maes, *Facts of Matter and Light*,
https://doi.org/10.1007/978-3-031-33334-7_5

> The whole cultural configuration at the turn of the century was in a change from mechanical to electromagnetic thinking. The immaterial electromagnetic concepts were to the same extent attractive as that the inert material images of the mechanics became unpleasant.
> — R. McCormmach, 1970.

Yet, the typical physicist around 1900 was not metaphysically unsettled by problems with the theoretical foundations of the profession. The vast majority were experimenters working on methods for measuring electrical effects, or studying heat phenomena, spectroscopy, or electrical discharges in gases.

There were problems, but there were also discoveries. During the last decades of the nineteenth century a number of new discoveries were made that urgently needed to be incorporated into the existing physics. At the end of 1895, Wilhelm Röntgen (1845–1923) discovered the famous X-rays. Radioactivity was discovered a year later by Henri Becquerel (1852–1908). And there were the discoveries of "discreteness'. The year 1897 is the official "birthyear" of the electron. It was already suspected from observation of the so-called Zeeman effect, where the influence of a magnetic field is to split lines in the spectrum of light emitted by certain gases. The electron was finally identified by Joseph Thomson (1856–1940) and immediately elevated to the universal building block of all chemical elements.

Important also was the development of an electron theory for metals. Paul Drude (1863–1906) was the pioneer here. He combined the study of optical, electrical, and thermal properties of metals. Drude and others used kinetic gas theory to explain electrical and thermal conductivity. Among other things, Einstein was struck by a paper written in 1901 by Max Reinganum, in which he described the electron theory as the basis for conductivity in metals. In it, Reinganum had written:

> It therefore turns out [...] that electricity also moves in metals in discrete quantities with the size of electrolytic ions, and that the principles of gas theory are applicable to the masses that move with the charges.

In 1916, Richard Tolman (1881–1948) and Thomas Dale Stewart (1890–1958) demonstrated experimentally that electricity consists of electrons flowing through a metallic conductor. They were able to measure the mass of the electron. As we repeatedly see, experiments bring us numbers.

5.1   Standard Hydrogen Lines

A very large part of what we know about Nature has been obtained through the experimental methods of spectroscopy and calorimetry, which center around the relations between heat, radiation, and the organization or structure of matter and energy. Those are the shores on which quantum mechanics was born. While spectroscopy and calorimetry have been complementary, and even mutually inspiring, the experiments with radiation were the ones that started the revolution. We adopt the same course here, and take light to be the main initiator.

In spectroscopy, we seek to measure radiation intensity as function of wavelength. In other words, we learn physics from colors, and still do so today: the intensity of a color produced by a fluid tells us about its nature, its internal dynamics, and its temperature. Much is technology-driven there. Think about a good light bulb: we want more light and less heat to come from it. This type of question was at the heart of experimental efforts in the second half of the nineteenth century.

When atoms get excited and couple to an electromagnetic field, they emit light. For example, a hydrogen gas, when placed in a tube with electric discharges, will start to emit light. In fact, understanding the spectrum of hydrogen was the analogue of something like deciphering the Rosetta stone. It contributed greatly to revealing the structures of the different atoms, establishing Mendeleev's periodic table, and going some way in explaining the interaction of light with matter.

Well before the advent of quantum mechanics, experiments showed that the emitted light consisted of a discrete set of frequencies. For the hydrogen spectrum, we usually refer to the discoveries and experiments by Lyman, Balmer, and Paschen. Observations were made using a spectrograph, an instrument which basically separates the incoming light (from a hydrogen discharge lamp) into different wavelengths. Transitions at these different wavelengths then appear as distinct lines in the spectrum. However, there is also Doppler broadening due to the thermal motion of atoms in the sample, which leads to lines turning into bell curves and spreading in frequency according to the formula

$$\Delta \nu \approx \frac{3}{2} \nu_0 \frac{k_B T}{M c^2} \, ,$$

where ν_0 is the center frequency, T is the absolute temperature, and M is the mass of the atom.

The wavelengths in the Lyman series lie in the ultraviolet band Lyman (1906, 1914). The Lyman-alpha line, denoted by Ly–α for a one-electron atom, is shown on the left of Fig. 5.1 with other lines discovered during the period 1906–1914 by Theodore Lyman (1874–1954) in the ultraviolet spectrum of an electrically excited hydrogen gas.

Johann Jakob Balmer (1825–1898) used visible light wavelengths between 400 and 700 nm Balmer (1885). In 1885, he described the known lines in the hydrogen spectrum by the formula

$$\lambda = 3646 \left(\frac{n^2}{n^2 - 4} \right) 10^{-10} \, \text{m} \, , \quad n = 3, 4, 5, \ldots \, ,$$

with the famous and brightest deep-red line Hα at about 656 nm (transition from $n = 3$ to $n = 2$). Friedrich Paschen (1865–1947) did the same with wavelengths in the infrared band Paschen (1908). See Fig. 5.1 for a summary.

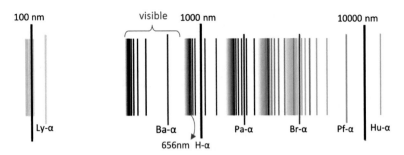

Fig. 5.1 Spectral lines of hydrogen, divided into series and shown on a logarithmic scale. The brightest hydrogen line in the visible spectral range is the deep–red H–α line, in the Balmer series with a wavelength of about 656 nm

In fact, as early as 1889, Johannes Rydberg (1854–1919) had already found a way to summarize the wavelengths of spectral lines as a generalization of the Balmer series Rydberg (1889). The Rydberg formula reads

$$\frac{1}{\lambda} = RZ^2 \left(\frac{1}{n_1^2} - \frac{1}{n_2^2} \right) , \quad n_1 < n_2 , \tag{5.1}$$

with λ the wavelength (in vacuum) of the emitted light, R the Rydberg constant, Z the number of protons in the atomic nucleus, n_1 the principal quantum number of the upper energy level, and n_2 the principal quantum number of the lower energy level of the atomic electron transition. Figure 5.1 indicates the α-lines, where $n_2 = n_1 + 1$. This formula applies to hydrogen-like chemical elements, where only one electron is affected by the effective nuclear charge.

In a nutshell, this showed that the lines (the various frequencies) form a pattern—the separations between them follow a specific mathematical rule. Clearly, this pattern had to be related to the structure of the atom. In fact, with the rebirth of the atomic hypothesis came the question of what an atom looks like. A first idea was that electrons should be spread throughout the atomic matter. Yet, Rutherford's experiments (see Sect. 7.2.1) showed that the atom had to have a nucleus. The atom could then be imagined as consisting of a positively charged nucleus surrounded by negatively charged electrons.

In 1913, Niels Bohr (1885–1962) proposed an *ad hoc* scheme much inspired by classical (solar) mechanics, where quantization schemes are applied to classical formulæ. The mathematics can be followed easily. Imagine a nucleus with charge Ze (where Z is the number of protons) and an electron with mass m and charge $-e$ in circular orbit with radius r around it. For mechanical stability, the centripetal force must equal the Coulomb force, so

$$\frac{Ze^2}{r^2} = \frac{mv^2}{r} .$$

Like planets moving around the Sun, electrons are also imagined to move in a plane with conserved angular momentum. Bohr assumed that

$$mvr = n\frac{h}{2\pi} \, ,$$

where n is an integer multiplying Planck's constant h divided by 2π.[1] This quantization rule may appear rather strange,[2] but let us see what it implies.[3]

First of all, as the product mvr can only take a certain sequence of values, it implies that not all orbits will be possible; they are quantized at certain distances from the nucleus. The details of moves from one orbit to another are not described, but at the same time, the electron orbit is somehow defined; see Fig. 5.2. Solving for v, we get

$$Ze^2 = mrv^2 = \frac{mrn^2h^2}{4\pi^2m^2r^2} = \frac{n^2h^2}{4\pi^2mr} \, ,$$

and so

$$r = \frac{n^2h^2}{4\pi^2mZe^2} \, , \quad v = \frac{2\pi Ze^2}{nh} \, , \quad n = 1, 2, \ldots \, .$$

The total energy of the electron is the kinetic energy plus the potential energy, i.e.,

$$E = \frac{mv^2}{2} - \frac{Ze^2}{r} = -\frac{mv^2}{2} \, ,$$

so the energy levels are

$$E_n = -\frac{2\pi^2mZ^2e^4}{n^2h^2} \, , \quad n = 1, 2, \ldots \, ,$$

working in units where $4\pi\epsilon_0 = 1$. We see that the quantization of the angular momentum has led to the quantization of energy. Taking $E = hc/\lambda$ for the radiated energy, we find the Rydberg constant

$$R = \frac{me^4}{8\epsilon_0^2h^3c} \, .$$

---

[1] John William Nicholson (1881–1955) had already introduced angular quantization in units of $\hbar$ in 1912, as quoted by Bohr in his 1913 paper. The notation $\hbar = h/2\pi$ was introduced by Paul Dirac in 1926.

[2] The so-called Bohr–Sommerfeld–Wilson quantization was closely related to the theory of adiabatic invariants in classical mechanics. Use of the principle of adiabatic invariance to define quantum numbers was introduced by Lorentz ("How does a simple pendulum behave when the length of the suspending thread is gradually shortened?") and Einstein ("both energy and frequency of the quantum pendulum changes such that their ratio is constant, so that the pendulum is in the same quantum state as the initial state.") at the 1911 Solvay conference.

[3] The discovery of electron spin, in the wake of the Stern–Gerlach experiment, which was motivated by the apparent strangeness of the Bohr–Sommerfeld quantization of the direction of the angular momentum, is discussed in Sect. 5.6.

Fig. 5.2 Some sketches of
the supposed orbits of
electrons in the
Bohr–Sommerfeld quantum
theory of various atoms,
from H.A. Kramers and
Helge Holst, *The Atom and
the Bohr Theory of Its
Structure* (New York: Alfred
A. Knopf, 1926), picture
from AIP

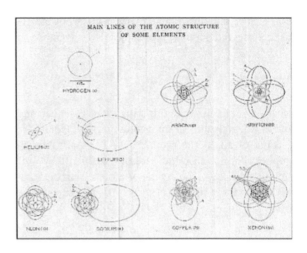

To be more precise, we should have used the reduced mass of the electron and nucleus. In 1916, Arnold Sommerfeld (1868–1951) applied relativistic corrections and examined the possibility of elliptical orbits.

If an electron jumps between orbits, it will emit the energy difference corresponding to the two quantum numbers n. This seems to go in the direction of an explanation for the particular form of lines in the hydrogen spectrum. All that constituted a major and indeed celebrated result: the phenomenological Rydberg formula (5.1) was obtained by applying quantization rules to classical mechanics. The rest of the story introduced so-called "quantum jumps."

Obviously, the ingenious derivation which starts by quantizing some sort of planetary motion around the nucleus is rather naive and possibly misleading (see Fig. 5.2). Today, sadly perhaps, many of us still carry around that visualization of the atom with electrons circling the nucleus, as shown in Fig. 5.2. That need not be the correct picture, of course.

Furthermore, such a hydrogen atom in thermal equilibrium is unstable, i.e., in a finite temperature and zero pressure environment. In large volumes and for entropic reasons, the atom spontaneously ionizes.[4]

The next step[5] (in 1926) was the creation of wave mechanics and the "solution" for the hydrogen atom by Erwin Schrödinger (1887–1961). This was improved upon in 1928 by Paul Dirac, who presented a relativistic quantum theory for the hydrogen atom. This theory predicted

---

[4] The thermal ionization formula carries the name of Meghnad Saha (1893–1956). It is of crucial importance for the interpretation of stellar spectra. Another application is for understanding the recombination of electrons and protons in the early universe.

[5] We forget the matrix mechanics introduced by Heisenberg (1925) and the corresponding calculation of the hydrogen atom by Pauli. Einstein's comment on matrix mechanics: "Heisenberg has laid a big quantum egg. In Göttingen they believe in it (I don't) [...] A veritable witches' multiplication table [...] exceedingly clever and because of its great complexity, safe against refutation."

$$E_{nj} = E_n \left[1 + \left(\frac{\alpha}{n}\right)^2 \left(\frac{n}{j+1/2} - \frac{3}{4}\right) \right], \qquad (5.2)$$

for the (modified) energy levels. The integer $n = 1, 2.3, \ldots$ is still the principal quantum number as in Bohr's formula, but now there is a correction involving the fine-structure constant

$$\alpha = \frac{e^2}{4\pi\epsilon_0\hbar c} \approx \frac{1}{137}.$$

The symbol j in (5.2) stands for the total angular momentum of the electron, combining the orbital and the spin angular momenta:

$$j = 1/2, \, 3/2, \, \ldots, \, n - 1/2.$$

We shall come back to this formulation when speaking about the Lamb shift (in Sect. 9.2). It was in fact only with quantum electrodynamics and again the pioneering work of Paul Dirac in 1931 that the interaction between light and matter got a fuller description.

All things considered, quantization of the hydrogen spectrum added a corpuscular aspect to radiation phenomena. The Lyman–Balmer–Paschen lines are obviously not the only spectroscopic signals from hydrogen atoms. Perhaps even much more important for the future of physics was the so-called 21cm line (microwave region), predicted by Henk van de Hulst (1918–2000), since this line gives important information about the early universe. First detected in 1951, it soon gave maps of the neutral hydrogen content of the Milky Way, and became an important tool for radio astronomy. It originates in the hyperfine transition in the ground state of neutral hydrogen, where the interaction between the magnetic dipole moments of the electron and proton spins results in a small increase in the energy when the spins are parallel. That spin-flip transition between the parallel and antiparallel states releases an energy of 2.874 µeV, corresponding to a wavelength of 21.106 cm. As such it is portrayed on the Pioneer plaque, attached to the Pioneer 10 and Pioneer 11 spacecrafts.

5.2 Black Body Radiation

One of the most remarkable aspects of the first quantum revolution (before 1930, say) was the way observation and experiment pointed quite unambiguously to what we know today as quantization and quantum discontinuity Norton (1993). Black body radiation was a prime example, but other phenomena all suggested that, with the aid of minimal assumptions, quantum discontinuity would be the unavoidable conclusion. It was a natural second step, so to speak, in the atomistic view of Nature. To follow the advice of Mark Kac (1914–1984): "Be wise, discretize."

Heating an object creates colors. We can explore that phenomenon more systematically for objects that are very good emitters of light, so-called black bodies. But perfect emitters are perfect absorbers, as they are each other under time-reversal: the emissivity is equal to the absorptivity for radiators in thermal equilibrium. The idea of a *black body* was first defined by Gustav Robert Kirchhoff (1824–1871) in 1859 as an object that absorbs all radiation falling upon it.

While the idea of a black body was invented by Kirchhoff, it still took some time to actually "make one," at least to a good approximation. One possibility was bodies that emit and absorb heat radiation within an opaque enclosure or cavity. Imagine something like an oven, but a rather impractical one—there is one small hole. We bring that oven to an equilibrium temperature T and we look at the radiation that comes out of the hole: what is the (volume or) intensity of the outgoing light as a function of the color. As Kirchhoff found from an application of the second law of thermodynamics, that spectrum would only depend on the temperature, not on the material details of the black body. Indeed, as previous experiments on radiant heat by Balfour Stewart (1828–1887) had also shown around 1855, the light that comes from such a black body has universal properties.[6] The plot of the intensity versus frequency appeared to be universal, only depending on the temperature T. Since color is really related to the frequency of the light, we get plots of the intensity of the black body radiation as function of frequency, for each temperature. That is black body radiation. For example, the Sun's radiation can be seen to a good approximation to come from a black body with an effective temperature of about 5800 K. Its emission peaks in the yellow-green part of the visible spectrum (not forgetting that there is a significant power output in the ultraviolet region as well, which is damaging to life).

Around 1890, the Physikalisch-Technische Reichsanstalt in Berlin was doing experimental research on thermal radiation. The focus was on applied physics and technology. The reason that lab scientists were looking into issues of thermal radiation was that they hoped to design a better electric light. In particular, there was a question of whether electricity or gas would be more economical for street lighting in Berlin. Hence the desire to develop a more precise standard for luminous intensity.

For that purpose, and for the practical generation of thermal radiation, various researchers were involved in developing the first cavity radiator. Most groundbreaking here were the experiments of Otto Lummer (1860–1925) and Alfred Pringsheim (1850–1941), and of Heinrich Rubens (1865–1922) and Ferdinand Kurlbaum (1857–1927) Lummer and Pringsheim (1899, 1901). Similar results were obtained by Paschen Paschen (1896) in 1900. They carried out measurements on black bodies and compared results with theoretical predictions. The latter, based on a formula by Rayleigh and Jeans, revealed serious problems with the high-frequency part of the radiation; this was the famous *ultraviolet catastrophe*, which predicted that almost all radiated power would be in the high-frequency domain (see Fig. 5.3). Experimental observation did not confirm this.

All this experimental activity was accompanied by the theoretical work of Wilhelm Wien (1864–1928) in 1896, and by people like Max Thiesen and Lord Rayleigh

---

[6] It is perhaps useful not to think of radiation as a surface phenomenon; the interior of the body is equally involved and we are dealing with a thermodynamic problem of radiation.

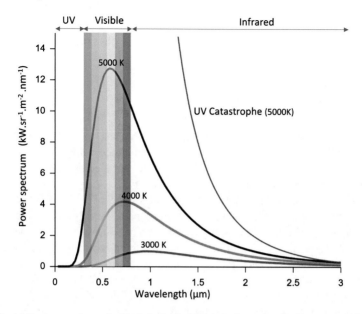

Fig. 5.3 As the temperature of a black body decreases, its intensity also decreases and its peak moves to longer wavelengths. Shown for comparison is the classical Rayleigh–Jeans law and its ultraviolet catastrophe

(1842–1919). It was Max Planck[7] (1900) who eventually discovered a way out, by postulating a quantization of the energy. In this way, he did away with the continuum aspect of electromagnetic fields.[8] The new theoretical formula fitted the experimental curve very well with just one fitting parameter, the so-called Planck constant h, which has dimensions of action, that is, energy × time, or distance × momentum. This new constant of Nature would survive all further developments of quantum theory, providing a measure of the distance between the quantum world and the classical world.[9] The idea that energy would come in the form of "particles" was perfectly

---

[7] The book *Black-Body Theory and the Quantum Discontinuity, 1894–1912* from T.S. Kuhn in 1978 is an extensive study of Planck's research program and the history of the first quantum theory Kuhn (1987).

[8] The history, motivations, and inspirations, are obviously much more complicated than this. For example, it appears that Planck did not accept Boltzmann's statistical interpretation of the second law of thermodynamics. He wished to show how irreversible processes could result from continuous matter, and the physics of black body radiation became his battleground.

[9] The use of the letter h for Planck's constant appears to be going back to notation used by the highly influential mathematician Augustin-Louis Cauchy, who was among the first to rigorously define the concept of limit and derivative. He used h whenever a small number was needed in some limiting procedure.

compatible with Boltzmann's view on entropy (and in particular, it resolved the problem of the ultraviolet catastrophe), but for most physicists (including Planck himself) it was still too hard to believe.

5.3   Photoelectric Effect

When we shine light on a metal plate, we can free electrons from the material surface. See Fig. 5.4. That requires a certain energy, the so-called work function or escape potential. The electrons that are released constitute an electric current that can be measured. Each electron has a certain kinetic energy.

Of course, a lot depends on the material, but it appears that if we start with red light (low frequency), the electrons may not escape at all, even if we increase the intensity of that red light. This is strange if we think of light as a wave, because strong waves would break any wall. But it turns out that there is a minimal frequency and it is only above this frequency that electrons are able to escape from the material. So what did not work for red light does work for blue or violet light. As we further increase the intensity of the light, more electrons get free but their individual kinetic energy does not increase. So what is happening?

It was Hertz who discovered the effect in 1887 Hertz (1887). The initial discovery was somewhat fortuitous. He was working on radio waves, one of the more sensational consequences Maxwell's electromagnetic theory of light. He had two metal plates (electrodes) a small distance apart, and applied a voltage across them. We call this a spark gap, because for a high enough voltage, sparking takes place. To see the spark, it helps to place the plates in a dark box. Hertz now observed that shining ultraviolet light on the two metal electrodes changed the voltage required for sparking to take place. This could be seen by changing the distance between the plates: in a dark box, the spark length was reduced, while in a glass box, the spark length increased. Putting the plates in a quartz box, the spark length increased still further. This was the first observation of the photoelectric effect. There appeared to be a direct relation between the light and the static electricity.

Fig. 5.4 The emission of electrons from a metal plate caused by light quanta, or photons

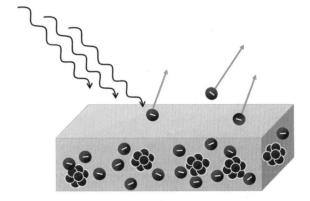

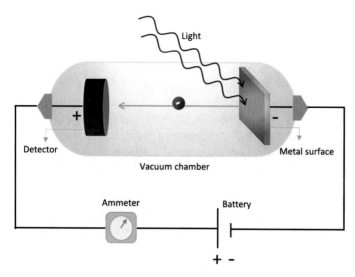

Fig. 5.5 Basics of the experimental setup for the photo-electric effect

Ironically, while Hertz was originally setting up experiments to prove Maxwell right, the effect he discovered could not be explained by Maxwell's electromagnetic theory: there, the kinetic energy should depend only on the light intensity and not the frequency.

The next step came a year later, with Willhelm Hallwachs (1859–1922), and this marked the beginning of systematic experimentation. Hallwachs connected a zinc plate to a battery and observed that shining ultraviolet (UV) light generated a current. The current was produced by the emission of electrons. Note that electrons were discovered by Joseph John Thomson (1856–1940) only in 1897. In 1898, that same J.J. Thompson found that the amount of *photoelectric* current varied with the intensity and frequency of the radiation used. Thompson's setup is still the standard way to illustrate the photoelectric effect; see Fig. 5.5.

We take two zinc plates (the electrodes) and place them at the opposite ends of a glass tube. A vacuum is maintained inside the tube. One of the electrodes (the cathode) is exposed via a small quartz window; ordinary glass blocks UV light. With a battery, we can vary the voltage across the two electrodes. We just need a battery and a potentiometer. We change the potential and the light intensity and we measure the current in the circuit. In the same vein, in 1902, Philipp Lenard (1862–1947) observed that the kinetic energy of the emitted electrons increased with the frequency of the radiation used.

Let us look a bit more closely at the standard theory to understand what is measured and how relevant that is. We are interested in the photoelectric current, which is the rate of emission of electrons. That current is directly proportional to the intensity of light falling on the electrode. So we will measure that, with increasing intensity, the current increases. Then, there is a voltage, and as it decreases the current also decreases. Yet, to obtain zero current we must reverse the voltage to the so-called stopping potential V_0 where the electrons can no longer quite reach the anode: therefore, the maximum kinetic energy of an emitted electron is

$$K = eV_0 \ .$$

The stopping potential V_0 is independent of the light intensity. Instead, that maximum kinetic energy increases with the frequency of the light, and the stopping potential gets more negative, i.e., the kinetic energy of electrons also increases. In fact, only light above a certain frequency ($\nu > \nu_0$) produces a photoelectric current. That frequency ν_0 is a threshold frequency that depends on the electrode material. Another observation is that the maximum kinetic energy of the electrons increases linearly with increasing light frequency. The minimum energy required for emission of an electron is called the work function of the material. We also observe that the electron emission occurs instantly, without any time lag.

The above observations were highly disturbing for physicists at that time. They seemed to contradict the energy concept in Maxwell's theory. The resolution would only come a few years later thanks to Einstein, who provided an explanation for the photoelectric effect Einstein (1905, 1906). Einstein had the idea of thinking of monochromatic light as consisting of particles (light particles, later called photons) that each have an energy proportional to the frequency, i.e., $E = h\nu$, where E is the energy and ν the frequency. Each photon can free an electron, but it needs sufficient energy, hence a sufficiently high frequency. This explains the phenomenon of the photoelectric effect.

That light consists of particles was an old idea; Newton had already entertained that idea, thinking of light as granular, with different particles for different colors. Yet, many experiments involving diffraction and interference had convinced physicists that light had to be a wave (see Chap. 6). Declaring that light consisted of particles looked therefore like a serious challenge and triggered many debates about the wave versus particle properties of light.

Let us therefore revisit Einstein's initial arguments in what is perhaps the most revolutionary of his 1905 papers. Einstein's approach differed greatly from that of Planck. He concentrated on the experimental part of the radiation law that was in strong conflict with classical calculations. The origin of the problem was the classical association of two quite different physical entities: the particle on the one hand and the electromagnetic field on the other. The latter has many more—in fact a continuum of—degrees of freedom. For equilibrium and equipartition, all excitations enter the field.[10]

When $h\nu/k_B T \gg 1$, i.e., for very high frequencies, we get Wien's law

$$e(T, \nu) \approx \frac{8\pi h\nu^3}{c^3} e^{-h\nu/k_B T} , \tag{5.3}$$

---

[10] The role of thermodynamics in Einstein's thinking is summarized in Klein (1967).

for the energy density at temperature T and frequency ν. Einstein expressed the total entropy

$$S_{\text{tot}} = V \int_0^{+\infty} s(\nu, e) \, d\nu \tag{5.4}$$

as an integral over entropy densities corresponding to frequency ν and energy density $e = e(T, \nu)$. We can compute that entropy density from the thermodynamic relation

$$\frac{1}{T} = \frac{\Delta s(\nu, e)}{\Delta e(T, \nu)} . \tag{5.5}$$

The result is

$$s(\nu, e) = -\frac{k_B e}{h\nu} \left(\log \frac{c^3 e}{4h\nu^3} - 1 \right) . \tag{5.6}$$

Let us now work monochromatically. The entropy in volume V is then $S = Vs$, where

$$S = S(\nu, E, V) = -\frac{k_B E}{h\nu} \left(\log \frac{c^3 E}{4h\nu^3 V} - 1 \right) ,$$

and $E = E(T, \nu) = Ve$ is the energy in the cavity at frequency ν. Comparing that with the entropy for a subvolume V_0, we get the difference

$$S - S_0 = \frac{k_B E}{h\nu} \log \frac{V}{V_0} . \tag{5.7}$$

Einstein noted that, for N particles in a volume V, the (ideal gas) entropy is $S = k_B N \log V$, and for a change $V_0 \to V$,

$$S - S_0 = N k_B \log \frac{V}{V_0} . \tag{5.8}$$

He concluded that, for monochromatic radiation, the entropy would vary with the volume like the entropy of an ideal gas, provided we identify

$$E = Nh\nu .$$

Here, N is now the number of light quanta with frequency ν.

The properties Einstein deduced were tested by Robert Andrews Millikan (1868–1953) in 1914 Millikan (1914, 1916).

The light particles would later be called *photons*, from a letter written towards the end of 1926 by Gilbert Lewis (1875–1946), *speaking of carriers of radiant energy*:

I therefore take the liberty of proposing for this hypothetical new atom, which is not light but plays an essential part in every process of radiation, the name photon.

5.4   Compton Effect

Perhaps the most spectacular demonstration that light manifests corpuscular properties goes under the name of the Compton effect. Arthur Holly Compton (1892–1962)
worked in Chicago and in Washington University in St Louis (Missouri). There, he
performed his most famous experiment, showing that light can be usefully represented as a flow of particles (photons) with momentum and energy determined by
the light wave's frequency.

In 1922, Compton found that X-rays could lose energy. They shift to longer
wavelengths after collisions with electrons, Compton (1923). See Fig. 5.6. The shift
in wavelength $\lambda' - \lambda$ and the scattering angle θ of the X-rays are related if each
scattered X-ray photon interacts with only one electron:

$$\lambda' - \lambda = \frac{h}{m_e c}(1 - \cos\theta) , \tag{5.9}$$

where the prefactor (involving the electron rest mass m_e) is called the Compton
wavelength of the electron, equal to about 2.43×10^{-12} m. In Compton's experiment,
some X-rays got no wavelength shift although they were scattered through large
angles anyway. That happens for "coherent" scattering off the atom as a whole,
where no electron is ejected. The Compton wavelength of the atom as a whole is
some 10 000 times smaller, which explains why no shift is visible then for the X-rays.

Compton attributed a particle-like momentum to the photons, in addition to the
energy that Einstein had attributed to it to explain the photoelectric effect (see
Fig. 5.6). The Compton formula (5.9) follows from conservation of energy and
conservation of momentum in the collision between the photon and the electron.

> This remarkable agreement between our formulas and the experiments can leave but little
> doubt that the scattering of X-rays is a quantum phenomenon.
> — A. Compton in the conclusion of his 1923 paper Compton (1923).

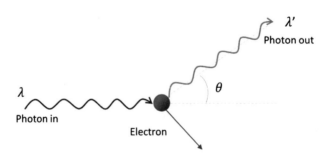

Fig. 5.6 A photon of wavelength λ collides with a target at rest. A new photon of wavelength λ'
emerges at an angle θ. The target recoils, carrying away an angle-dependent amount of the incident
energy

The Compton effect, or Compton scattering, demonstrated once more the particle-like nature of electromagnetic radiation. The principle of energy momentum applies strictly and not just statistically.[11] The relevant length scale is the Compton wave length $\lambda = h/mc$. The classical theory using Thomson scattering tried to explain the shift in the wavelength as radiation-pressure recoil with an associated Doppler shift of the scattered light. The idea was that, for sufficient intensity of the electric field, the associated light will accelerate charged particles to a relativistic speed. Yet that effect would become negligibly small at very low light intensities, independently of wavelength.

Note that Compton collisions dominate when the photon energy becomes large compared to the binding energy that holds the electron in an atom. In the original experiment by Compton, the energy of the X-ray photon was about 17 keV, larger than the binding energy of the atomic electron. Therefore, the electrons could be treated as free, in contrast to the situation in the photoelectric effect. The target was carbon. For carbon, the Compton effect prevails over the photoelectric effect above 20 keV, for copper above 130 keV, and for lead above 600 keV. After the collision, the (high-speed) electron loses its energy by ionization as a beta-electron, but the scattered photon propagates through the material, interacting again and undergoing multiple Compton scattering until it is absorbed as in the photoelectric effect. On the other hand, when the photon energy exceeds 1 MeV, a new phenomenon arises in the transformation of the photon into an electron and a positron. This phenomenon becomes prominent, for example, in particle accelerators.

The Compton effect is responsible for an important interaction of high-energy X-rays with atomic nuclei in living beings and as such is applied in radiation therapy. Inverse Compton scattering occurs when the photon gains energy, or the charged particle transfers part of its energy to the photon. It is important in astrophysics and plasma physics. For example, low-energy photons produced from thermal processes may be scattered to higher energies by relativistic electrons in stellar atmospheres or plasma. The Sunyaev–Zel'dovich effect is another instance: when photons from the cosmic microwave background move through a hot gas, they are scattered to higher energies. This thus provides a method for detecting galaxy clusters. The mathematics which governs the transition between photon frequencies when photons undergo Compton scattering is summarized in the Kompaneets equation (1956) Kompaneets (1957); Freire Oliveira et al. (2021). It uses a quantum version of the Boltzmann equation and describes the relaxation to the Planck distribution as a Fokker–Plank equation (which becomes nonlinear due to stimulated emission) Freire Oliveira et al. (2022).

---

[11] That was confirmed by an experiment by Walther Bothe and Hans Geiger in 1924 via coincidence measurements using a Geiger counter.

5.5 Specific Heat

Apart from the radiation phenomena, a second basic reason through which it became clear that the understanding of physics was incomplete in the nineteenth century involves the specific heats of gases. It was the second cloud Lord Kelvin mentioned in his 1900 speech; see the quote from Lord Kelvin at the beginning of this chapter.

The nineteenth century brought with it the kinetic gas theory (or molecular theory), and hence also the beginnings of statistical mechanics. Even though it met with many successes, there was also a great puzzle: the theory could not account for the behavior of heat capacities, especially at low temperatures. Indeed, experiment contradicted theory. The theory applied the so-called equipartition of energy, which gave a clear prediction of specific heats that was just not compatible with experiment results (see Fig. 5.7). As Maxwell had already concluded:

> I have now put before you what I consider to be the greatest difficulty yet encountered by the molecular theory.
> — J.C. Maxwell, 1860s.

Maxwell was not the only one to be concerned; James Jeans (in his *Dynamical Theory of Gases*, 1904) and Josiah Willard Gibbs (in his *Statistical Mechanics*, 1902) were also greatly disturbed by the situation.

Calorimetric measurements by Dulong (1785–1838) and Petit (1791–1820) had shown (1819) that specific heats of solids had a common value (see Fig. 5.7). Atoms in the solid were taken to be simple harmonic oscillators about equilibrium positions, with three independent vibrations, so that the total energy of one mole of the solid comes out equal to $3N_A RT$. The molar heat capacity would be $3R$, i.e., about 6 calories per degree.

However, for some elements, and in particular the lighter ones, such as beryllium, boron, and carbon, it was found that that they had much lower heat capacities than

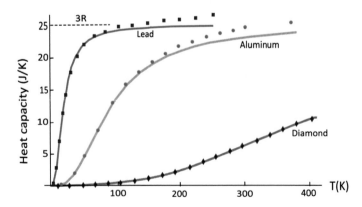

Fig. 5.7 Heat capacities of some solids and their (different) low-temperature values, showing their deviation from the Dulong–Petit law

the Dulong–Petit value, and they varied rapidly with temperature to approach the Dulong–Petit value only at temperatures well above room temperature (see Fig. 5.7). Moreover, at the beginning of the twentieth century, evidence was accumulating that atoms had an internal structure; for example, the presence of electrons meant that there could be electronic vibrations.

When Einstein first came to the Solvay Conference in Brussels in 1911, this was his subject: "On the Present State of the Problem of Specific Heats." It is essential once again to recall that quantum theory was to a great extent developed by dedicated efforts to carry out difficult and precise experiments. As already discussed, there were the experiments on radiation that were discussed in Sect. 5.2. But there was also the subtle work done to measure the specific heat of molecular hydrogen and other diatomic gases at low temperatures. For example, these studies demonstrated the freezing out of the rotational degrees of freedom. As a matter of fact, the low-temperature measurements of the specific heat of solids and liquids by Walther Nernst (1864–1941) and for liquids by Frederick Alexander Lindemann (1886–1957) were crucially important in opening the way to a quantum theory of matter (and not only radiation).

The theoretical question was formulated succinctly by Einstein in his report to the First Solvay conference, when he referred to :

> [...] the highly important but, unfortunately, mainly unsolved question: How is mechanics to be reformulated so that it does justice to the radiation formula as well as the thermal properties of matter?

There was indeed a good reason why the scientists of the nineteenth century got the heat capacity wrong; while the theorists of matter were all eventually atomists, they did not know about the quantization of energy. Everything is related to quantization and the fundamental discreteness of degrees of freedom. Some of the main ideas were in that way inspired by the desire to unify, as Einstein put it:

> I now believe that we should not be satisfied with this result. For the following question forces itself upon us: If the elementary oscillators that are used in the theory of the energy exchange between radiation and matter cannot be interpreted in the sense of the present kinetic-molecular theory, must we not also modify the theory for the other oscillators that are used in the molecular theory of heat? There is no doubt about the answer, in my opinion. If Planck's theory of radiation strikes to the heart of the matter, then we must also expect to find contradictions between the present kinetic-molecular theory and experiment in other areas of the theory of heat, contradictions that can be resolved by the route just traced. In my opinion this is actually the case, as I try to show in what follows.
> — Einstein in his report at the first Solvay Conference.

The contradictions that Einstein mentions refer to the violations of the equipartition theorem.

Einstein began to resolve these contradictions by stressing the universality of the quantum hypothesis: "If Planck's theory strikes to the heart of the matter", then the average energy and thus the specific heat of any oscillator must depend on the frequencies of the atomic vibrations in the solid. Assuming for simplicity, as Einstein did, that all atomic vibrations are independent and of the same frequency v, the energy U of 1 mole of the solid becomes

$$U = 3R\frac{h\nu}{k_B}\frac{1}{\exp(h\nu/k_B T) - 1},$$

and the specific heat follows by differentiating this with respect to temperature. It is easy to see that the heat capacity is negligibly small when $k_B T/h\nu < 0.1$ and takes the equipartition value when $k_B T/h\nu \gg 1$. Light atoms could be expected to vibrate at higher frequencies than heavier ones, and this already gave a qualitative insight into why the light elements had anomalously low heat capacities.

Nowadays, we still use (equilibrium) calorimetry to probe the energy occupation levels of materials. Getting these right[12] was among the greatest successes of the quantum revolution and of the idea of quantization as universal mechanism.

5.6 Spin

And now for a strange story. The Stern–Gerlach experiment Gerlach and Stern (1922); Schmidt-Böcking et al. (2016) of 1922 set out to test the hypothesis of directional quantization, as introduced by Bohr (1913) in his atomic model, and extended by Sommerfeld (1916) (see Sect. 5.1).

Otto Stern (1888–1969), whose basic training had been in soft condensed matter physics, found it quite ridiculous to suggest that the angular momentum of electrons in their motion around the nucleus could be quantized. This would somehow mean that atoms were oriented:

> If this nonsense of Bohr should in the end prove to be right, we will quit physics!
> — Oath sworn by Otto Stern and Max von Laue.

Stern found Walther Gerlach (1889–1979) ready and capable to do a "magnetic" experiment. The setup is well known (see Fig. 5.8).

A narrow beam of silver atoms, obtained by evaporation from an oven heated to a temperature of about 1300°C, pass a second aperture that selects those atoms with velocity parallel to the y-axis. Before crossing the electromagnet, the magnetic moments of the silver atoms are oriented randomly. After leaving the region of the nonuniform magnetic field, the particles hit a screen and, not without a certain drama and some serious experimental effort, were observed to move in two possible directions. Classically, as for spinning magnetic dipoles, a random and continuous spot would have been seen. Hence the idea that "Bohr was right after all" (see the postcard in Fig. 5.9). The title of the Stern–Gerlach paper Gerlach and Stern (1922) was "Experimental evidence of directional quantization in a magnetic field."

A strange story indeed: silver has no orbital angular momentum. What they measured was spin, and their experiment was going to become paradigmatic for quantum measurements in general. When teaching measurement procedures or the Copen-

---

[12] It was Peter Debye (1884–1966) who produced a solid-state equivalent of Planck's law of black body photon radiation, getting, e.g., the correct T^3 law for the low-temperature specific heat.

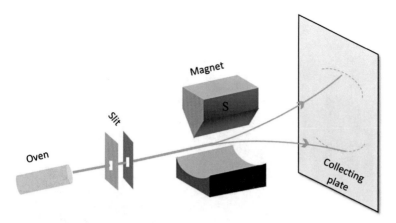

Fig. 5.8 Setup of the (traditional) Stern–Gerlach experiment

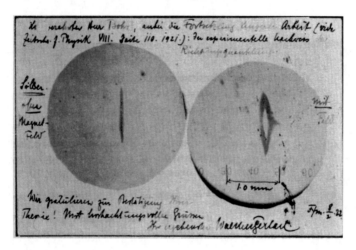

Fig. 5.9 Success at last! Postcard sent to Bohr. The excitement was great: "Bohr is right after all." (W. Gerlach to O. Stern). "This should convert even the nonbeliever Stern" (W. Pauli to W. Gerlach)

hagen interpretation of quantum mechanics, many textbooks refer to the setup of the Stern–Gerlach experiment.

Yet, spin is really an intrinsic form of angular momentum, a "hidden" rotational degree of freedom. "Perhaps there exists a quantized rotation of the electrons," Compton had written, Compton (1921). This idea was first proposed by Wolfgang Pauli in 1924 Pauli (1925), calling it a fundamental *Zweideutigkeit*, and it would eventually give rise to the Pauli exclusion principle. At the beginning of 1925, Ralph Kronig (1904–1995), PhD student at Columbia University, proposed electron spin after hearing Pauli, who understood it as a "ridiculous" rotation of the electron in space: "it is indeed very clever but of course has nothing to do with reality." Kronig decided not to publish his theory, but the fact remains that Kronig had told Pauli

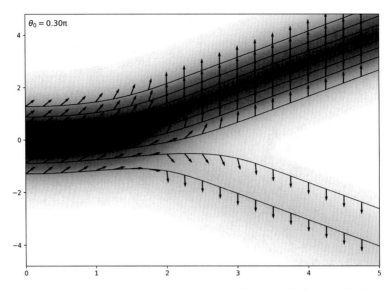

Fig. 5.10 Simulation of the Stern–Gerlach trajectories with initial spin direction biased upwards. Courtesy Simon Krekels

about electron spin before Pauli published "On the Connexion between the Completion of Electron Groups in an Atom with the Complex Structure of Spectra" in 1925 Pauli (1925). A couple of months later, George Eugene Uhlenbeck (1900–1988) and Samuel Abraham Goudsmit (1902–1978), students of Paul Ehrenfest (1880–1933), took it in a somewhat different direction, but were able to explain the Zeeman effect Uhlenbeck and Goudsmit (1926). One can also conclude here that Uhlenbeck and Goudsmit were the first to introduce electron spin, as it seemed to confirm what was said in Pauli's paper. In the end, "spin" is a fundamental quantum property of the wave function used to describe particles. Yet, the Stern–Gerlach experiment was the first evidence for spin, as explained by Ehrenfest in 1927. See also Pais (1989) for more history and references.

Using the Dirac–Pauli equation, the influence of spin on quantum trajectories can be simulated and visualized (see Fig. 5.10 and Gondran and Gondran 2016; Maes et al. 2022).

Chapter 6
Wave-like Nature

The Ancient Greeks had explained vision as consisting of visual rays extending between the eye and an object. In that sense it is quite normal to perceive light as consisting of particles. After all, one of the first things that catches the eye is that light propagates in straight lines. Rays are straight. We learn that lesson from early on; the game of hide-and-seek is based on it. Newton for sure got convinced that light consists of particles, like little bullets moving fast and straight:

> *Query 28.* Are not all Hypotheses erroneous, in which Light is supposed to consist in Pression or Motion, propagated through a fluid Medium? For in all these Hypotheses the Phænomena of Light have been hitherto explain'd by supposing that they arise from new Modifications on the Rays; which is an erroneous Supposition.
> — Isaac Newton, Opticks (published in English in 1704)

Could it be that Newton foresaw problems with the æther as a medium for light? His corpuscular theory of light was not at all the one we have today, but it was very elaborate nevertheless.

Newton thought that light consisted of corpuscules, little bullets, each with its own color. The refraction of light at a water surface could be reduced to collisions of those light particles with that surface. In this way, he was able to explain the appearance of the rainbow. Spectra originate by filtering out the particles that wear the appropriate color. White light consists of all kinds of color particles, and when the light beam passes through a prism (or water droplet), depending on the color, slightly different forces work so that the beam breaks up into a color spectrum.

Not everyone followed Newton. Famously, Christiaan Huygens (1629–1695) thought of light as a wave motion. Various experiments were performed that showed that it is actually not quite true that light propagates perfectly rectilinearly. Depending on the dimensions of holes and corners (when comparable with the wavelength of light), we can see light bending; that is diffraction.

Interference and diffraction belong to the world of waves, effects we are used to with sound waves and water waves; see Fig. 6.1. Diffraction is the bending of waves around the corners of an obstacle or through an aperture into the region of the geometrical shadow. Interference is the combination of two or more waves by adding their displacement together at every spacetime point, to form a resultant wave of lower, higher, or the same amplitude. For example, interference patterns of sound waves can be heard by placing two sound sources at about 1m apart in a large open space and setting each to emit the same frequency; distinct areas of loudness and softness will be observed. On the other hand, diffraction is caused by the interference of many secondary wavelets that originate from a single wave, making the difference between maxima and minima much poorer.

In 1660 Francesco Grimaldi (1618–1663) was already making precise observations of light diffraction. The findings of Grimaldi were demonstrated in two experiments. In the first, Grimaldi let a beam of sunlight enter a darkened room through a very small hole. He found that shadows cast by small opaque bodies were larger than they ought to be if light traveled in straight lines. In the second experiment, he found that, when two very small apertures were placed one after another, light passing through both cast a spot larger than any geometric theory of optics could explain. It was as if the supposed pencil of light took on the shape of a cone.

Interference of light was first studied in the double-slit experiments of Thomas Young (1773–1829) around 1802. This contributed significantly to the idea that light is a wave. We can indeed easily explain the diffraction and interference of light if we think of it as a wave. Some 60 years after Young's experiment, Maxwell discovered that the repeated (or sequential) application of his third and fourth laws provided a mechanism for the generation and propagation of an electromagnetic wave. He soon realized that the wave could be identified with light. Theory and experiment agreed.

So, what a surprise it was to find in the twentieth century that matter could also show interference and diffraction. As they move, constrained by the geometry, elementary units of matter may together build interference and diffraction patterns upon impacting a screen. The trajectories of matter may be so non-Newtonian that diffraction and interference patterns form. Indeed, as the de Broglie wavelength h/p gets sufficiently large compared to the relevant geometry, the motion of elementary particles like electrons appears to be guided by waves in configuration space. Very visible in the motivation and inspiration of Erwin Schrödinger in 1926, following ideas put forward in the 1924 Ph.D. thesis by Louis de Broglie (1892–1987), optics and mechanics found their new unified empirical basis in slit experiments, and this then developed into quantum mechanics.

6.1 Early Light

6.1.1 Young's Experiment

Despite Newton's enormous academic weight, the idea that light did not simply consist of particles moving in straight lines gained traction in the nineteenth century. The first experiments that truly started changing people's minds were suggested by Thomas Young:

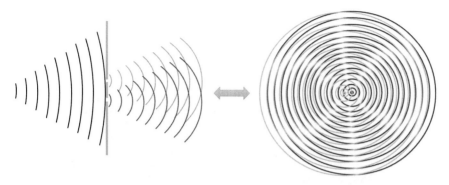

Fig. 6.1 Interference of water waves passing through two slits

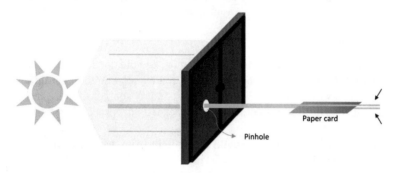

Fig. 6.2 Creating a double source by placing a card in the sunlight beam

The experiments I am about to relate [...] may be repeated with great ease, whenever the sun shines, and without any other apparatus than is at hand to every one.
— Thomas Young, Bakerian Lecture: Experiments and calculations relative to physical optics. Philosophical Transactions of the Royal Society **94**, 1–2, 1804.

In 1801, Young presented "On the Theory of Light and Colours" to the Royal Society. He organized for a horizontal beam of sunlight, entering the room through a little hole in the window shutter. He placed a card measuring approximately 0.85 mm in a beam of sunlight, Fig. 6.2. That gives two coherent sources and the light beams from the both sides of the card interfere. One could see the fringes of color in the shadow and to the sides of the card. He observed that placing another card in front of or behind the narrow strip so as to prevent the light beam from striking one of its edges caused the fringes to disappear. This was certainly evidence that light is composed of waves.

More famous, however, is Young's interference experiment with a double-slit, which crucially supported the undulatory theory of light (see Fig. 6.1). While there is actually no clear evidence that Young carried out the two-slit experiment, he did describe it in some detail:

In order that the effects of two portions of light may be thus combined, it is necessary that they be derived from the same origin, and that they arrive at the same point by different paths, in directions not much deviating from each other. This deviation may be produced in one or both of the portions by diffraction, by reflection, by refraction, or by any of these effects combined; but the simplest case appears to be, when a beam of homogeneous light falls on a screen in which there are two very small holes or slits, which may be considered as centres of divergence, from whence the light is diffracted in every direction. In this case, when the two newly formed beams are received on a surface placed so as to intercept them, their light is divided by dark stripes into portions nearly equal, but becoming wider as the surface is more remote from the apertures, so as to subtend very nearly equal angles from the apertures at all distances, and wider also in the same proportion as the apertures are closer to each other. The middle of the two portions is always light, and the bright stripes on each side are at such distances, that the light coming to them from one of the apertures, must have passed through a longer space than that which comes from the other, by an interval which is equal to the breadth of one, two, three, or more of the supposed undulations, while the intervening dark spaces correspond to a difference of half a supposed undulation, of one and a half, of two and a half, or more.
— Thomas Young

Young was perfectly aware of what was going on, and did quantitative analysis from the observed patterns. Interference was produced by sending the light through two small holes in a wall. The two slits were the secondary sources of light, and Christiaan Huygens had described their propagation, exactly as for all other waves.

He thought about the possible paths of the light from the two point sources to a given point. When the difference in path lengths is an integer number of wavelengths, the two waves add together to give a maximum in the brightness, whereas when the path difference is equal to half a wavelength, or one and a half, etc., then the two waves cancel and the intensity is at a minimum. Young used this calculation to estimate the wavelengths of violet and red light.

Of course, it helps to use monochromatic light and Young only had sunlight to work with. However, white light consists of different colors. For every color, the interference pattern has a maximum in the center, while the spacing varies with the wavelength. It is a nice feature that the superimposed patterns therefore vary in color, with maxima in different places depending on the color. Nevertheless, only two or three fringes are easy to observe.

The conclusion was that light shows wave properties, quite contrary to what Newton and the optical atomists had in mind, so it seemed. Naturally, from classical mechanics which was the only mechanics imagined at that time, these would move in straight lines and just show two little dots on the screen, one behind each opening. What Young saw instead was that the intensity of light on the screen was minimal in exactly those regions on the screen; much more light was falling on the center of the screen, midway between the two slits. The whole pattern could be explained by exploiting the idea of superposed waves. The result was then an interference pattern on the screen behind, made up of streaks of light and darkness.

6.1.2 On the French Side

The story goes on. The wave theory of light was becoming more and more widely accepted, but the debate was not over. A memorable incident happened in connection with a thesis by Augustin-Jean Fresnel (1788–1827), submitted to the French Academy of Sciences in 1818 in answer to a call for papers challenging the corpuscular theory of light. Fresnel explained Huygens' principle and Young's principle of interference. However, Siméon-Denis Poisson (1781–1840), one of the referees, thought he had found an error, mainly because he himself was convinced of the corpuscular nature of light. He argued that, if Fresnel and the wave theory were correct, there would have to be a bright spot in the middle of the shadow of a circular obstacle blocking a point source of light. Poisson was sure that was absurd. But the experiment had to be done, and it was Dominique François Jean Arago (1786–1853) who took up that challenge.

Arago observed what is now called the Arago spot, i.e., the bright spot predicted by Poisson, and which seemed to him so *absurd*; see Fig. 6.3. This was another decisive step in convincing scientists, now on the French side, of the wave nature of light. In the end, Fresnel won the challenge. As a matter of fact, and as Arago later noted, "his" spot had been observed before by Joseph-Nicolas Delisle (1688–1768) and Giacomo Filippo Maraldi (1665–1729) a century earlier in Paris.

Although the observation of the Arago spot did not settle the debate completely, the final blow to the corpuscular theory came only a few decades later. To explain Snell's law of optical refraction, Newton had to assume that light propagates faster in denser media, contrary to the prediction of the wave hypothesis. Of course, while Newton had some idea of the speed of light c due to astronomical observations by Ole Roemer (1644–1710), and Huygens actually calculated c from Roemer's data, no experiment would be capable of measuring the difference for almost two centuries. Arago himself, still a believer in the particle nature of light, had declared in 1838

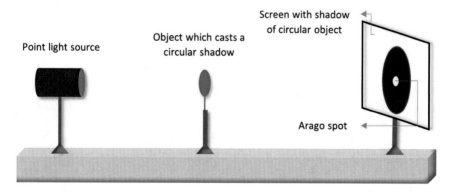

Fig. 6.3 Arago spot experiment. A point source illuminates a circular object, casting a shadow on a screen. At the shadow's center, at least in the so-called Fresnel regime, a bright spot appears due to diffraction, contradicting the prediction of geometric optics

that one or the other theory would succumb the day when, by direct experience, we would find out whether light traveled faster in air or water. It was around that time that the speed of light was first directly measured by Hyppolyte Fizeau (1819–1896).

Fizeau's apparatus consisted of a cogwheel, rotating several hundred times per second. A ray of light was aimed at a remote mirror, through the teeth around the edge of the wheel. By adjusting the rate of rotation, there comes a point where the light going through one notch of the wheel is eclipsed by the next tooth moving into place during the round trip. Knowing the distance to the mirror, one can then work out the speed of light, which he determined to be 70 948 "leagues of 25 per degree," or 315 576 km/s.

When made aware of Arago's proposal, both Fizeau and Léon Foucault (1819–1868) raced to complete this *experimentum crucis*, with Foucault beating Fizeau to the punch by a few weeks. Foucault improved on the Fizeau apparatus by removing the cogwheel and bouncing the light toward the remote mirror with another, rapidly rotating mirror instead. Any tiny rotation of this mirror occurring during the back-and-forth motion of the light would result in a displacement of the reflected image, and this would become measurable after magnification by a telescope. Most likely due to the fact that there was no practical way to make a long straight water container measuring a few kilometers long, Foucault's result was only qualitative, but it nevertheless conclusively proved that light travels more slowly in water, driving the final nail in the coffin of the particle theory of light.

It remains to explain what actually makes light travel more slowly in water, and how that relates to the wave *versus* particle picture of light. First and briefly, light always travels at the same speed really: the change in speed is only apparent, not factual, even when it is made to slow down enormously.[1] It is not at all like a bullet being slowed down by collisions in the medium. It is due rather to interference. The particles in the medium (atoms) begin to produce electromagnetic waves of their own via all the oscillating electrons. More and more bouncing waves appear and when all these waves are added up, we obtain a refractive index giving a speed about 25% lower in water.

6.1.3 Interacting Newton Bullets?

The mere existence of diffraction and interference were not sufficient to convince everybody of the wave nature of light as well. After all, the corpuscles were supposed to be overwhelmingly numerous. Who was to say it was not possible that, due to some extremely weak interactions between them, interference might still arise as an emergent property. Perhaps one may stubbornly think that diffraction and interference are the result of interactions between particles. Or even that classical (Newtonian) mechanics applies to these particles with very small mutual interactions able to

---

[1] Experiments in the group led by Lene Hau have managed to slow light down to very low speeds, including stopping it completely, only to release it afterwards Hau et al. (1999, 2001).

produce wave phenomena, or as the result of local interactions with an æther. Some people do indeed believe this.

With this question in mind, a low-intensity double-slit experiment was first performed in 1909 by Geoffrey Taylor (1886–1975) Taylor (1909). To create a low-intensity source, he simply placed smoked glass screens between the source and the detection screen. He took five photographic images, each exposed to the same amount of light, steadily diminishing the intensity of the light. The longest exposure took three months to complete. He found that:

> In no case was there any diminution in the sharpness of the pattern […].

It is indeed important to know that interference of visible light produces fringes even with extremely weak light sources. For example, the interference effects with light do not result from a direct interaction, e.g., one photon interfering with another photon. Each photon makes a small dot on the screen. The interference pattern is built up (independently) photon by photon. The experiment was redone in 1986 using a single-photon source, and the same result was obtained Grangier et al. (1986).

Triple-slit Young experiments have now been performed. One of the challenges there is to avoid misalignment of the slits, which might produce errors. Urbasi Sinha and colleagues carved into a metal coating on a glass plate three equally spaced transparent slits, 30 μm wide and 100 μm apart. A laser fired photons one at a time at the plate and the result was the wavy interference pattern we all expected Sinha et al. (2010).

The above does not exclude that "other" mechanics might produce interference patterns with (noninteracting) bullets. Indeed, quantum mechanics is very different. There we get matter waves, and while those are guided by wave functions and formally share some of the light properties, we must not confuse waves with particles.[2] The (non-Newtonian) trajectory picture of quantum particles is fully compatible with wave-like detection outcomes and phenomena, and is in fact one of the biggest, "more recent," and useful visualizations of quantum mechanics Bohm (1953), Goldstein (1998), Gondran et al. (2005), Maes et al. (2022).

6.2 X-rays and Bragg Scattering

The X in X-rays stands for their *unknown* origin, at least in the understanding of Wilhelm Röntgen (1845–1923), who discovered them in the autumn of 1895, and also of the many people before Röntgen who had observed this unidentified radiation emanating from experimental discharge tubes. Röntgen was also experimenting with tubes and accelerating electrons, and he wrote an initial report entitled "On a new

---

[2] Forget wave-particle duality. Wave mechanics is not in general used as a synonym for the mechanics of quantum particles, as described by Bohmian mechanics, Bohm et al. (1952). Bohmian mechanics describes motion of matter as guided by the wave function, and is named after David Bohm (1917–1992).

Fig. 6.4 The first published
X-ray diffraction pattern
along the fourfold axis of
sphalerite Friedrich et al.
(1912)

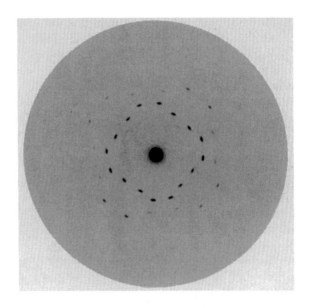

kind of ray: A preliminary communication." Submitted to the journal of Würzburg's
Physico-Medical Society, it was the first paper written on X-rays. Röntgen had immediately understood the possible medical applications, and they would come thick and
fast, especially when World War I broke out; even in 1914, Marie Curie was already
on the front line in Flanders with "radiological cars" to support injured soldiers.
In these vehicles, rapid X-ray imaging of wounded soldiers was possible, helping
surgeons to operate more accurately.

Research in physics also continued. The most important step was made by a group
working with Max von Laue (1879–1960). In 1912, Walter Friedrich (1883–1968)
and Paul Knipping (1883–1935) made the first observation of X-ray diffraction by
crystals Friedrich et al. (1912). A pattern of spots appeared, similar to the pattern
when visible light diffracts through holes or gratings (see Fig. 6.4). The experiments
came first and the theory followed. From the standard theory of (light) diffraction,
von Laue understood that the space within the regular structure of a crystal must
be similar in size (angstrom) to the wavelength of X-rays (ranging from 10 pm to
10 nm). These are the conditions under which diffraction occurs. The conclusion
was clear: X-rays were waves. It was a great experimental discovery, and it gave
birth to more discussion and exploration. It would also spark the beginning of X-ray
crystallography, important for physics and much beyond.

This next step was brought about by the studies of William and Lawrence Bragg,
father and son. In fact, the father, William Bragg (1862–1942), was a firm proponent
of the theory that X-rays are made up of particles. However, his son Lawrence (1890–
1971) was a research student at Cambridge at that crucial time. Being a student and
having just taken exams on light, crystals, and waves, Lawrence's mind was free
and fresh, and he had access to all the relevant theory. He got the idea that the

Fig. 6.5 Bragg diffraction. Two beams with identical wavelengths and phases approach a crystalline solid and are scattered off two different atoms within it. The lower beam traverses an extra length of $2d \sin \theta$. Constructive interference occurs when this length is equal to an integer multiple of the wavelength of the radiation

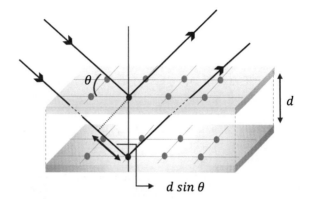

pattern of dots produced by passing X-rays through a crystal might be caused by a series of reflections. Some X-rays reflect off the first sheet of atoms, some from the second, and so on. He even managed to formulate the "Bragg equation," relating the wavelength λ of the X-rays, the distance d between the successive atomic sheets or layers in the crystal, and the angle θ at which the X-rays hit the sheets. This relation $n\lambda = 2d \sin \theta$, for some integer n, determines the value of θ at which the constructive interference is strongest (see Fig. 6.5).

The father William Bragg set up apparatus in which a crystal could be rotated to precise angles while measuring the energy of reflections. In this way, the distances between the atomic sheets could be measured in a number of simple crystals. Going the other way, they obtained the wavelengths of X-rays produced by different metallic targets in X-ray tubes from knowledge of the spacing of the atoms (calculated from the weight of the crystal and the Avogadro constant) Bragg and Bragg (1913).

Together with the work by von Laue mentioned above, these experiments founded the area of X-ray crystallography. Diffraction with X-rays became the tool of choice to analyze crystal structure. It was now possible to calculate the positions of atoms in crystals. Both Bragg and Laue diffraction deal with X-rays and are useful crystallographic techniques. Laue diffraction is more general, simply dealing with the scattering of waves by a crystal. Bragg diffraction gives the angles for coherent and incoherent scattering from a crystal lattice.

Lawrence Bragg became the successor of Ernest Rutherford as the Cavendish Professor of Experimental Physics in Cambridge. Years later (in 1951), X-ray crystallography would be applied by Rosalind Franklin (1920–1958) and Maurice Wilkins (1916–2004). It turned out to be of great importance to unravel the double helix structure of DNA. In 1953, James Watson and Francis Crick at the Cavendish began to build a molecular model of DNA and the data derived from Franklin's research helped to get it right.

6.3 Davisson–Germer Experiment

In the 1920s another diffraction experiment was performed, this time by Clinton Davisson (1881–1958) and Lester Germer (1896–1971). But it was not light in the form of X-rays or any other form of electromagnetic radiation that was scattered from a crystal, but matter particles, and in particular, electrons. These are considered to be structureless electric charges with a mass, quite different from light. And yet, a diffraction pattern was observed, similar to that for X-rays. This showed that the electron, and in fact "matter" in general, can exhibit wave-like properties under certain circumstances.

The original inspiration was the work of Louis Victor de Broglie (1892–1987). His PhD thesis Broglie (1924) "Recherches sur la théorie des quanta" in 1924 postulated wave properties for electrons and for all matter. We have here the basis of wave mechanics, as generalized in the work of Erwin Schrödinger (1926) and later reformulated in the work Bohm et al. (1952) of David Bohm (1952). It was strongly supported by Einstein, and indeed confirmed by the electron diffraction experiments we discuss here.

For the history, let us first meet Walter Elsasser (1904–1991). He started as a PhD student in Göttingen early in 1925. He regularly attended the "Structure of Matter" seminars organized there by Max Born (1882–1970). One of the earliest presentations in this seminar was by Friedrich Hund (1896–1997), a student of Born. He is remembered in Hund's rules and for his contributions to the theory of atomic spectra. The report in the seminar was about an experiment by Davisson and Kunsman at Bell Telephone Laboratories. It was a scattering experiment in which electrons were shot at a platinum plate. They found that the intensity of the distribution of the electrons varied with the scattering angle Davisson and Kunsman (1922, 1923). There were maxima and minima for different values of the scattering angle, which was all rather mysterious. This should be compared with the Rutherford experiments of 1911 (see Sect. 7.2.1). In the abstract of their paper Davisson and Kunsman (1922, 1923), Davisson and Kunsman wrote:

> Since low speed electrons are more easily deflected than α-particles, their scattering patterns depend not only on the field immediately about the nucleus but also upon its nature in regions beyond various or all of the structural electrons.

They thought there were shells of electrons at certain distances from the nucleus and that they were responsible for the array of maxima and minima as a function of the scattering angle. Born also tried to explain the variable deflection minima and maxima by shells of electrons that were of different densities. Whether this suggestion was correct or not, nobody knew.

Remember, however, that much of traditional scattering theory was already formulated in the 1920s. For the collision of an electron with an atom, Born used the ideas of wave scattering, as dictated by the correspondence between particles and their guiding de Broglie waves. It was in the same paper that Born first introduced the notion of probability into quantum mechanics. It was very natural then; the wave function was a statistical guide for the particles in the sense that the amplitude of the wave specified a probability for the particle to travel in certain ways.

It was in this spirit that Walter Elsasser was pursuing his studies. Visiting the library one day in May 1925, he found two recent papers by Einstein on a remarkable quantum effect for gases, later to be called the phenomenon of Bose–Einstein condensation. One way to understand that heuristically is to think of the collection of gas particles as sitting "under the same wave." At certain values of the parameters specifying the density and temperature, the particles show mutual coherence, and they behave together in a stable condensate. The relevant length scale (to be compared with the density) is the thermal wavelength $h/\sqrt{2\pi m k_B T}$ for massive particles. Einstein was inspired here by the PhD thesis of Louis de Broglie, which Elsasser also found in the university library. The thesis contained de Broglie's basic idea that matter is guided by a wave, at least at the appropriate scales: the wavelength λ is related to the particle's momentum by the de Broglie formula $\lambda = h/p$. And here Walter Elsasser wondered: could it not be that Davisson and Kunsman's maxima and minima were diffraction phenomena, similar to those produced by X-rays penetrating crystals? Here the diffraction would be produced by a slight penetration and reflection of the electrons. Elsasser easily calculated the energy of the electrons required for the maxima, and it came out just right, as needed for understanding the experimental results of Davisson and Kunsman Davisson and Kunsman (1922, 1923). This became a short note, positively reviewed by Einstein and reproduced by Max von Laue in 1944 in a book on matter waves.

The experiments of Davisson and Germer spanned the period 1923–1927 and were performed again at Western Electric (Bell Labs) in New Jersey. In the summer of 1926, Davisson attended the Oxford meeting of the British Association for the Advancement of Science where he heard from the thoughts and the work of Elsasser and Born. He went back home to improve upon the experimental setup, with the great results as reported in Davisson and Germer (1927, 1928).

The experiment was carried out with a cubic nickel crystal (see Fig. 6.6). In fact, the real and initial aim of the experiment was to study the surface of nickel. The atoms are symmetrically arranged there, in planes parallel to the end surfaces of the crystal. In other words, the atoms form a quadratic network in the planes. If, however, an angle of the cube is symmetrically cut off, the radiation surface becomes a triangular plane, with the atoms in this plane forming a triangular network.

You take a piece of stone,
chisel it with blood,
grind it with Homer's eye,
burnish it with beams
until the cube comes out perfect.

Next you endlessly kiss the cube
with your mouth, with others' mouths,
and, most important, with infanta's mouth.
Then you take a hammer
and suddenly knock a corner off.

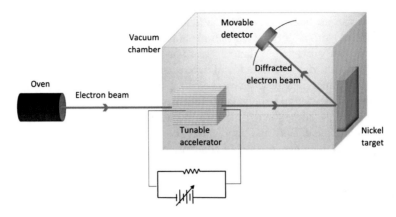

Fig. 6.6 Experimental setup of the Davisson–Germer experiment Davisson and Germer (1927, 1928)

All, indeed absolutely all will say
what a perfect cube this would have been
if not for the broken corner.
— Nichita Stanescu (1933–1983), Lecture on the Cube.

Davisson and Germer sent a tiny bundle of electrons of some fixed velocity to hit the triangular plane perpendicularly. The speeds were low, obtained by passing an electron through a voltage of between 50 and 600 V. Suppose we had waves instead of incoming electrons: the wave plane would be parallel to the surface of the crystal, striking the atoms on the surface simultaneously. Each atom would be the center of a new wave moving in all directions, and we want to think about waves having the same effect as the corresponding electrons. The detector would capture the waves going out parallel to the crystal plane and at right angles to one of the sides of the triangle. Parallel to that side, the atoms lie in parallel rows with a certain distance between the rows. That distance was determined before, using X-rays. Every row emits its wave, but surely the waves from the inner rows must arrive later: they must travel a longer distance to reach the edge of the triangle. At any rate, some irregular system of waves is emitted in which waves interfere, neutralizing each other when no outgoing wave is produced. If the waves have such a wavelength that the distance between the rows of atoms becomes equal to a multiple of the wavelength, then all the outgoing waves will be in phase and they will reinforce one another. Via de Broglie's relation, the velocities of the electrons correspond to wavelengths, and hence we can calculate for what velocities their wavelength equals the distance between the rows of atoms. Davisson found that the theory agreed very well with the experiment results.

There was no longer any doubt. In the following years, the investigations met with even greater success: the existence of matter waves and the correctness of de Broglie's theory was established. Diffraction of electrons was a milestone in the understanding of quantum mechanics.

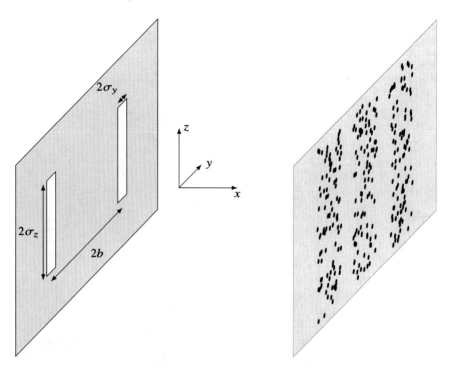

Fig. 6.7 Cartoon of the spatial setup used in Maes et al. (2022), with electrons exiting the wall from slits to arrive at the screen

Four months later, as reported in Thomson (1927), George Paget Thomson (1892–1975) used much faster electrons, accelerated by voltages between 10 000 and 80 000 V. Such swift electrons were subsequently of great use to study the structure of matter. Thomson used another setup from Davisson and Germer. He used very thin films of celluloid, gold, platinum, or aluminum, and the electron beam was made to fall perpendicularly upon the film. He then examined the diffraction figures produced on a fluorescent screen placed behind the film. The thickness of the films was 10^{-7}–10^{-8} nm, still containing many small crystals in various directions. On the screen, Thomson saw a series of concentric rings. The theory proposed by de Broglie told him that they would correspond to the various directions of the crystal planes. Moreover, the diameter of a ring would give the wavelength of the matter wave which agreed with the spacing of the planes. It was a method used by Debye and Scherrer to study the structure of crystals using X-ray analysis. Thomson found excellent agreement with de Broglie's theory. Rather nicely, when Thomson applied a magnetic field, it influenced the beams (as they were electron beams), and he observed a lateral movement of the image on the screen. Whereas his father (J.J. Thomson) had seen the electron as a particle, Thomson the son demonstrated that it could be diffracted like a wave.

Fig. 6.8 Simulation result
from Maes et al. (2022).
Near-field behavior of the
trajectories for a single slit in
the Schrödinger picture

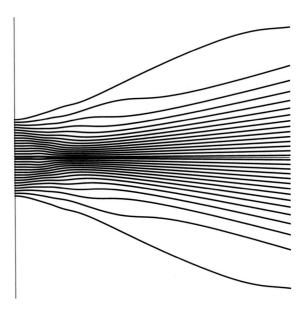

Subsequent experimental work also confirmed de Broglie's theory for beams of molecules, atoms, and atomic nuclei. In addition, looking back at the previous section, in 2009, the Arago spot was obtained using a supersonic expansion beam of deuterium molecules, which gave an example of neutral matter waves. Furthermore, an Arago spot could be seen using transmission electron microscopes for electrons diffracted from certain circular structures in ultrathin specimens. X-ray interferometry has a neutron counterpart. Neutron interferometry began in the 1970s with pioneering work by Helmut Rauch (1939–2019) and Samuel Werner Rauch and Werner (2015). In 1974, Rauch, Ulrich Bonse, and Wolfgang Treimer demonstrated the first interference of neutrons. For neutrons, the de Broglie wavelength is approximately 1000 times smaller than for an electron at the same energy.

6.4   Wavy Electrons

The Young experiment for electrons took long to establish. Clauss Jönsson (1930) was the first to do so, in 1961 Jönsson (1961, 1974). Jönsson's electrons had a de Broglie wavelength of 5 nm (to be compared with the wavelength of roughly 500 nm for visible light).

Pioneering work on electron interference was certainly also contained in Merli (1976), but the most sensational results were those published in 1989 by a team led by Akira Tonomura (1942–2012) at Hitachi Tonomura et al. (1989). They performed a double-slit experiment where the electrons arrived one by one. Naturally, each electron produced a little light dot on the screen. As more and more electrons were sent through, the pattern that accumulated had a form that was again typical of a wave with two centers at the openings in the wall. In these experiments, electrons are not smeared out across the screen but rather behave like point particles. Each electron strikes the screen at a different point, and an interference pattern emerges only when many electrons have hit the screen.

Also in the 1980s, Tonomura and his colleagues used a toroidal ferromagnet (a doughnut, six micrometers in diameter) at 5 K and covered it with a niobium superconductor to completely confine a magnetic field within (Meissner effect). The phase difference from the interference fringes was measured between one electron beam passing though the hole in the doughnut and the other passing around the outside of the doughnut. The interference fringes were displaced by just half a fringe of spacing inside and outside the doughnut, confirming the existence of the Aharonov–Bohm effect.

Today, we are also able to simulate the electron's motion (as a particle guided by the Dirac equation), to visualize its trajectories, and see how they make the diffraction and interference patterns on a screen placed opposite a wall with one or more slits Maes et al. (2022) (Fig. 6.7). The pioneering ideas are those of de Broglie and of Bohm. Bohm in particular wrote down the guiding equations, Bohm et al. (1952).

After the discussion with Einstein, I began to think on other lines, and I began to think that this Schrödinger equation resembled a Hamilton-Jacobi equation, and you could think of a particle being *effected* by a wave. That was the model which I proposed in papers in 1952.
— David Bohm, in the New Scientist Interview, 11 November 1982 (Fig. 6.8).

Visualizations and simulations are important for understanding the role of electron spin and to get a clear view of the arrival-time distribution. Those features are still very difficult to assess experimentally.

The wave function $\psi(\mathbf{x}, t) \in \mathbb{C}^4$ for a single electron in four spacetime dimensions satisfies

$$(i\hbar\gamma^\mu\partial_\mu - mc)\psi = 0 , \tag{6.1}$$

where we use the Dirac gamma-matrices γ^μ. In the so-called Weyl or chiral basis for the gamma-matrices, we can write the wave function as a bispinor, i.e., a pair of two-component Weyl spinors $\psi = (\psi_-, \psi_+)^\mathrm{T}$, corresponding to, respectively, a left-handed and a right-handed Weyl electron. On that basis, we rewrite the Dirac Eq. (6.1) as

$$i\hbar\,\sigma^\mu\partial_\mu\psi_+ = mc\psi_- , \quad i\hbar\,\overline{\sigma}^\mu\partial_\mu\psi_- = mc\psi_+ , \tag{6.2}$$

which are coupled equations for these spinors, with $\sigma^\mu = (1, \sigma)$ and $\overline{\sigma}^\mu = (1, -\sigma)$ for Pauli matrices σ. In the same way, the Dirac current

$$j^\mu = \overline{\psi}\gamma^\mu\psi , \quad \text{where} \quad \overline{\psi} = \psi^\dagger\gamma^0 , \quad \text{satisfying} \quad \partial_\mu j^\mu = 0 ,$$

can be decomposed as $j^\mu = j_+^\mu + j_-^\mu$ for

$$j_+^\mu = \psi_+^\dagger\sigma^\mu\psi_+ , \quad j_-^\mu = \psi_-^\dagger\overline{\sigma}^\mu\psi_- .$$

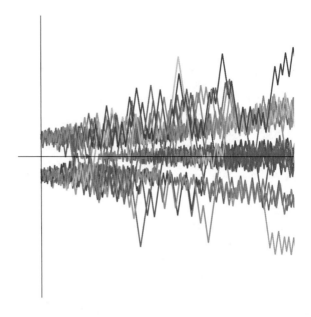

Fig. 6.9 Simulation result from Maes et al. (2022). Dirac–Pauli trajectories of zigzag electrons exiting a double slit

These each satisfy a continuity equation with a source

$$\partial_\mu j_\pm^\mu = \pm F \ , \quad \text{where} \quad F = 2\frac{mc}{\hbar}\mathrm{Im}\psi_+^\dagger\psi_- \ ,$$

which is exactly the same as

$$\partial_t\rho_+ + \nabla\cdot(\mathbf{v}_+\rho_+) = a\rho_- - b\rho_+ \ ,$$
$$\partial_t\rho_- + \nabla\cdot(\mathbf{v}_-\rho_-) = b\rho_+ - a\rho_- \ , \tag{6.3}$$

for the densities $\rho_+ = j_+^0 = \psi_+^\dagger\psi_+$ and $\rho_- = j_-^0 = \psi_-^\dagger\psi_-$, where

$$\mathbf{v}_+ = c\frac{\psi_+^\dagger\boldsymbol{\sigma}\psi_+}{\psi_+^\dagger\psi_+} \ , \quad \mathbf{v}_- = -c\frac{\psi_-^\dagger\boldsymbol{\sigma}\psi_-}{\psi_-^\dagger\psi_-} \ ,$$
$$a = 2\frac{mc^2}{\hbar}\frac{\left(\mathrm{Im}\psi_+^\dagger\psi_-\right)^+}{\psi_-^\dagger\psi_-} \ , \quad b = 2\frac{mc^2}{\hbar}\frac{\left(\mathrm{Im}\psi_-^\dagger\psi_+\right)^+}{\psi_+^\dagger\psi_+} \ , \tag{6.4}$$

for $F = F^+ - (-F)^+$ and $F^+ = \max\{F, 0\}$

We can now interpret (6.3) and (6.4) as the master equation for an ensemble of independent particles following

$$\dot{\mathbf{x}}_t = \mathbf{v}_{\chi_t}(\mathbf{x}_t, t) \ , \tag{6.5}$$

with positions $\mathbf{x}_t$ at time t. The variable $\chi_t = \pm$ is the chirality, where right-handed Weyl electrons have $\chi = +$ and left-handed electrons have $\chi = -$. This χ_t follows an inhomogeneous Markov jump process with transition rate $a(\mathbf{x}, t)$ to jump from $- \rightarrow +$ and rate $b(\mathbf{x}, t)$ to jump from $+ \rightarrow -$ when $\mathbf{x}_t = \mathbf{x}$. These give the stochastic trajectories of two different massless manifestations of the Dirac electron, "zig" and "zag," i.e., as the Weyl electrons Penrose (2005) (Fig. 6.9).

The speed $|\mathbf{v}_\pm| = c$ is always equal to the speed of light. At every transition $\chi \rightarrow -\chi$, the direction of the particle changes from $\mathbf{v}_\chi$ to $\mathbf{v}_{-\chi}$. The tumbling (and the change in the effective velocity field) is a consequence of the coupling between left- and right-chirality and occurs at a rate proportional to the mass. Or, *vice versa*, the mass is generated by the tumbling and the zig-zag motion is due to the fundamental massless nature of the particles. The resulting dynamics can be simulated as a time-dependent Markov process $(\mathbf{x}_t, \chi_t)$ and it reproduces at every moment the correct densities $\rho_\pm$, while remaining time-reversal invariant. For more details, see Maes et al. (2022).

Chapter 7
Finding Structure: Scattering and Fission

Molecules, atoms, and nuclei have structure. They make things and they are made up of things. To unravel their interiors, we throw things at them, light or heavy particles, and we see how they sing or sound, break or transform.

7.1 Light Scattering

Scattering experiments were the bread and butter of much of modern physics through-out the twentieth century. There are many types of scattering, however. Let us divide our account here into scattering by waves (such as light) and scattering by particles. Between those two the oldest subject of study is perhaps light scattering.

We may take the beginning of the concept of light scattering to fall in the seventeenth century with Christiaan Huygens. Remember his principle:

> Every point of material that is hit by light is initiated to emit light of the same wavelength.

Huygens' principle includes the idea that light is a wave. Waves also scatter from obstacles, and the way they do so depends on the size of the obstacles and the wavelength of the light.

The experimental work on light scattering by John Tyndall (1820–1893) takes us into the nineteenth century Tyndall (1869). He showed how we can see the trail of light in a slightly turbid colloidal solution. He studied the scattered light, and in particular, its polarization and the way it depends on the scattering angle. The motivation for the experiments was a simple worry: in 1869 Tyndall wanted to determine whether any contaminants remained in purified air. He discovered that bright light became faintly blue-tinted when scattered off nanosized particles. From this, he suggested that a similar scattering of sunlight gave the sky its blue hue.

C. Maes, *Facts of Matter and Light*,
https://doi.org/10.1007/978-3-031-33334-7_7

John William Strutt, better known as Lord Rayleigh (1842–1919), took it from there Rayleigh (1899, 1918), starting in 1870. Rayleigh scattering is the (predominantly) elastic scattering of electromagnetic radiation by particles much smaller in size than the wavelength of the radiation. It requires the particle to be less than approximately 3% of the light wavelength. Molecules of nitrogen and oxygen are a few thousand times smaller than the wavelength of visible light, so air is good and ready to scatter sunlight. This stems from the electric polarizability of atoms or molecules. Light has an oscillating electric field that acts on the charges within a particle. They start to move at the same frequency, which means that the particle becomes a small radiating dipole. The radiation is what we see as scattered light. We can see it when light travels through transparent solids and liquids, but it is most prominently seen in gases.

The amount of Rayleigh scattering is inversely proportional to the fourth power of the wavelength. Moreover, the scattered light comes out mostly at right angles to its original direction, with the bending being much stronger for blue light than for red light. Sunlight hits a gas molecule high in the atmosphere, and twists and turns in a zig-zag fashion down to the Earth's surface. Some light that has been through thousands of these scattering events ends up in our eyes. In that way, Rayleigh explained the blue color of the sky as due to elastic scattering of light by molecules in the atmosphere. Red light, seen in the morning and evening when the Sun is viewed through more of the atmosphere, is scattered much less. The only color light that is not scattered away on that long journey is the red. Blue light from the sun is scattered in all directions and much more so than the other colors. When, at the daytime, you look up at the sky, you see blue no matter where you look.

Scattering can occur directly from particles, but more generally also from density fluctuations. An example is critical opalescence, a phenomenon near a continuous or second-order phase transition. It was already reported on by Charles Cagniard de la Tour (1777–1859) in 1823 for mixtures of alcohol and water. It was also used by Thomas Andrews (1813–1885) in 1869 for his experiments on the liquid–gas transition in carbon dioxide. It is most commonly demonstrated in binary fluid mixtures: as the critical point is approached, the length scales over which the density fluctuates grows. The correlation length of the liquid diverges. When the size of density fluctuations becomes comparable with the light wavelength, scattering causes the normally transparent liquid to appear cloudy. In 1908, Marian Smoluchowski (1872–1917) understood that critical opalescence is due to large density fluctuations, while Einstein showed in 1910 the theoretical link with Rayleigh scattering. As a matter of fact, it is still extremely useful to read the first few pages of that Einstein paper, Einstein (1910), also in order to understand the practical relevance of the macroscopic fluctuation theory touched upon in Sect. 4.3. In a letter to Wien, then editor of Annalen der Physik, Einstein wrote: "Please accept [this manuscript] for the *Annalen*," and added:

> The first part of the paper, which deals with Boltzmann's principle, is perhaps too lengthy. But please do not take this amiss; I have been wanting to expound on my opinion about this topic, and this was an opportunity to do so.
> — A. Einstein, 7 October 1910.

The next step in the investigation of light scattering was to connect it with quantum physics. An early example here was the work by Chandrasekara Venkataraman Raman (1888–1970). His interest in waves and vibrations had developed in his early life from trying to understand musical instruments and the sounds they could make. It was at the Indian Association for the Cultivation of Science in Kolkata that he began his life as an experimenter. Raman did not ask why the sky is blue, but rather why the sea is blue, supposedly on a boat trip back from England in 1921. He got the idea that the striking blue color of the Mediterranean Sea is perhaps not simply a reflection of the sky. With a small spectroscope and a prism, he immediately began to investigate light scattering by the water and noted that water molecules could scatter light just as air molecules could. The result was the discovery in February 1928 of a new scattering phenomenon, the Raman effect.[1]

Recall what happens in Compton scattering (1922): X-rays can lose energy in collisions with electrons, thereby shifting to longer wavelengths. Raman made visible light scatter inelastically off molecules. or these investigations, still in Kolkata, Raman's team (with K.S. Krishnan in particular) started simple experiments to study light scattering in various liquids. Using colored filters, they tried to detect a visible change in color of the scattered light; this required an intense light source, for which Raman used a telescope, concentrating sunlight for the experiments. The group thus discovered a shift in the color of light scattered by many different liquids, and in 1928, Raman observed that the scattered light was polarized, which meant that this effect was not a kind of fluorescence.

In contrast with the elastic Rayleigh scattering, Raman scattering is an inelastic scattering process with a transfer of energy between the molecule and the scattered photon (see Fig. 7.1). It is much rarer, happening to approximately 1 in 10 million photons. Energy can be either gained or lost, but more frequent is the so-called Stokes–Raman scattering, where the molecule gains energy from the photon during the scattering (excited to a higher vibrational level) and the scattered photon loses energy and increases its wavelength. The reason why the ocean is blue is the same as why the sky is blue, but it has nothing to do with reflections. In addition, red and other long wavelength light are absorbed more strongly by water than is blue. Mostly the blue gets returned.

Others such as Grigory Landsberg (1890–1957) and Leonid Mandelstam (1879–1944) in 1925 looked at light scattering in quartz. In 1928, they independently observed the same scattering effect that Raman had found. Not long after, Raman scattering (also called, combinational scattering of light) acquired a central position in the experimental study of the vibration and rotation of molecules. We can say that the colors of the light that pass through a substance make its fingerprint.

---

[1] The phenomenon of Raman scattering of light was first postulated by Adolf Smekal in "Zur Quantentheorie der Dispersion," 1923.

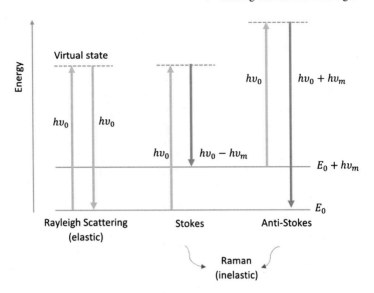

Fig. 7.1 Diagrammatic representation of an energy transfer model for Rayleigh scattering, Stokes–Raman scattering, and anti-Stokes–Raman scattering. Molecules in a liquid change the color of some of the light passing through it

7.2 Particle Scattering

From our discussion of the particle-like properties of light and radiation and the wave-like properties of matter, it appears somewhat arbitrary to insist on distinguishing particle and wave scattering. Concerning targets, they would be matter, obviously.

7.2.1 Geiger–Marsden–Rutherford Scattering

Particle physics came on the scene. It involved firing a beam at a target. The point was to shoot at things and see where the bullets would fly.

This was the beginning of the twentieth century, and while atoms had been hypothesized as the building blocks of matter, their structure remained unknown. It was the group led by Ernest Rutherford that made a fundamental experimental discovery. Some twenty years after the Rutherford experiment, Eddington would marvel at the result:

> The revelation by modern physics of the void within the atom is more disturbing than the revelation by astronomy of the immense void of interstellar space.
> — Sir Arthur Stanley Eddington In: The Nature of the Physical World (1928).

It is interesting to give some historical details here, if only to highlight the fact that the Rutherford experiment was not directly intended to discover the structure of the atom. Ernest Rutherford (1871–1937), known as the father of nuclear physics and for many in Great Britain considered to be the greatest experimentalist since Faraday, had already discovered the nature of α and β radiation. In fact, that was the real origin and motivation of the experiment.

Remember that α-particles are tiny, positively charged particles (helium nuclei). Their discovery, by Rutherford in 1899, had everything to do with radioactivity; α-particles are spontaneously emitted by elements like uranium and radium (discovered by Marie Curie, who was awarded the 1911 Nobel Prize in Chemistry for this). Rutherford was trying to make a precise measurement of their charge-to-mass ratio in 1908. The first thing was to count them, i.e., how many α-particles would a sample of radium yield. If the total charge was known, he could find the charge of one α-particle.

The first experiments were performed with a screen coated with zinc sulfide; there would be a flash of light each time an α-particle hit it. You can imagine Rutherford and his assistant Hans Geiger sitting in the dark and trying to count the flashes of light. But first, their eyes had to become sensitive enough to not miss those flashes from the ZnS screen.

It was probably the problem of sitting in the dark straining their eyes that made them think of another way of counting. This used the following idea: α-particles ionize air molecules, and if the air is within an electric field, an electric current will be produced by the ions. This inspired Rutherford and Geiger to create a counter which consisted of two electrodes in a glass tube: each α-particle that passed through the tube gave a pulse of electricity that could be counted. This was indeed the first Geiger counter.

The scattering experiment could now begin, under the direction of Rutherford at the Physical Laboratories of the University of Manchester. The target was often a gold foil. Because gold is unusually ductile, it can be hammered into fine sheets, to make a foil that is less than a half a micron thick. Think of the art in those days,[2] and in fact over many centuries: artists and artisans alike used such gold leaf to make their creations shimmer and shine, and add that ethereal touch.

Rutherford found that a narrow beam of α-particles was broadened when it passed through a thin gold film and Geiger measured the scattering angle for the outgoing α-particles after hitting the foil. Geiger found that the scattering was small, on the order of one degree. Now at this time, Ernest Marsden (1889–1970), a twenty year old undergraduate, was also working in Rutherford's lab. Geiger suggested that Marsden could carry out a research project, and Rutherford responded, "Why not let him see whether any α-particles can be scattered through a large angle?" Rutherford was always ready with a nice phrase; here is one for the management:

---

[2] E.g. in the Art Nouveau that spread from Belgium in the 1890's to the rest of Europe, or think of the Golden Phase of Gustav Klimt (1862–1918) in his *Portrait of Adele Bloch-Bauer I* (1907) and *The Kiss* (1907–08) (Fig. 7.2), perhaps inspired by Byzantine Imagery.

Fig. 7.2 The Kiss, by
Gustave Klimt in his Golden
Period. At the same time, the
Rutherford experiment firing
particles at atoms, remind us
of the words of W.H. Auden
writing about a universe
"Wherein a lover's kiss
Would either not be felt Or
break the loved one's neck"

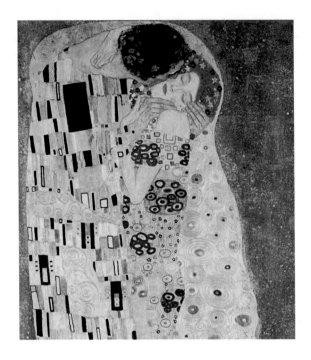

Every good laboratory consists of first rate men working in great harmony to insure the
progress of science; but down at the end of the hall is an unsociable, wrong-headed fellow
working on unprofitable lines, and in his hands lies the hope of discovery.
— E. Rutherford.

The work of Geiger and Marsden was beautiful and interesting also from the social
point of view. Geiger and Marsden worked in great harmony and were of course both
very interested in science and their experiments. They were, however, on opposite
sides in two wars. During World War I, Marsden served in France as a Royal Engineer,
while Geiger served in the German military as an artillery officer from 1914 to 1918.
Basically the same happened again during World War II, with Geiger organizing
scientific efforts to support the military in Germany, and Marsden in New Zealand.

In 1909, the atom was still imagined as a plum pudding, this being the model
proposed by the discoverer (and quantifier) of the electron, Joseph John Thomson
(1856–1940) in 1897. The plums would be the electrons (but moving rapidly), and
the pudding, giving the consistency, would be a positively charged spherical mass.
In that year, Geiger and Marsden went on to check. They took a piece of gold foil
and fired a beam of α-particles at it (Fig. 7.3).

And indeed, the experiment showed that the plum-pudding model was not correct:
the positive charge formed a concentrated nucleus which was able to exert large
Coulomb forces and thus deflect the α-particles through large angles as they passed
by. In Geiger and Marsden (1909), Geiger and Marsden described how α-particles
were sometimes scattered by more than $90°$.

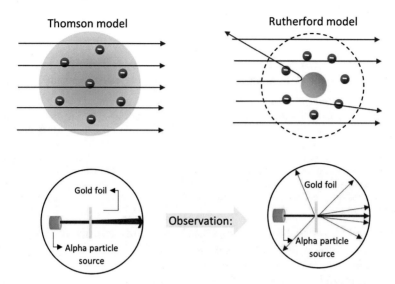

Fig. 7.3 *Left* Had Thomson's model been correct, all the α-particles should have passed through the foil with minimal scattering. *Right* What Geiger and Marsden observed was that a small fraction of the α-particles experienced strong deflection

It should be noted that the discovery of large deflections was not and indeed should not be taken immediately as saying anything about the atomic structure. The Geiger–Rutherford counter used for the detections was unreliable because the α-particles were also being strongly deflected by their collisions with the molecules of air inside the detection chamber. In fact, Rutherford thought that α-particles were too heavy to be strongly deflected. He asked Geiger to investigate how the properties of the target influenced the scattering: just how much matter was needed to scatter α-rays? Now came experiments to observe how the foil scattered α-particles for different thicknesses and different types of material.

In the same 1909 paper Geiger and Marsden (1909), Geiger and Marsden also noticed that metals with higher atomic mass, such as gold, reflected more α-particles than those with lower ones, like aluminum. Geiger went on to determine how the most probable angle through which an α-particle is deflected varies with the target material, its thickness, and the velocity of the α-particles. He could vary the velocity of the α-particles by placing extra sheets of mica or aluminum at the source. All this went into his 1910 paper, on "The scattering of the α-particles by matter". Geiger concluded Geiger (1910):

- the most probable angle of deflection increases with the thickness of the material,
- the most probable angle of deflection is proportional to the atomic mass of the substance,
- the most probable angle of deflection decreases with the velocity of the alpha-particles,

- the probability that a particle will be deflected by more than 90° is vanishingly small.

When Geiger reported to Rutherford that he had spotted α-particles being strongly deflected, Rutherford was astounded. In a lecture Rutherford delivered at Cambridge University, he said:

> It was quite the most incredible event that has ever happened to me in my life. It was almost as incredible as if you fired a 15-in. shell at a piece of tissue paper and it came back and hit you. On consideration, I realized that this scattering backward must be the result of a single collision, and when I made calculations I saw that it was impossible to get anything of that order of magnitude unless you took a system in which the greater part of the mass of the atom was concentrated in a minute nucleus. It was then that I had the idea of an atom with a minute massive centre, carrying a charge.
> — Ernest Rutherford

The key for discovering the nucleus was that, occasionally, an α-particle would enter into direct collision with the nucleus of one of the 2000 or so atoms it had to pass through, after which it would be scattered at much larger angles, sometimes fired right back toward the source.

It was time for Rutherford to come up with a mathematical model for the scattering pattern, that fitted the conclusions, not just qualitatively, but also quantitatively. This became the famous 1911 paper "The scattering of α and β particles by matter and the structure of the atom." There, Rutherford proposed his idea that the atom contains at its center a region of electric charge that is very small and intense Rutherford (1911). He assumed that this central charge was positive, but he could not prove that. The result depended on developing a mathematical equation for how the foil should scatter the α-particles if all the positive charge and most of the atomic mass were concentrated in a single point at the center of the atom. Rutherford went on to give a detailed study of the elastic scattering of charged particles by the Coulomb interaction Rutherford (1911).

This gave rise to the planetary model of the atom, later also called the Bohr model, as discussed in Sect. 5.1. It explained how something like 1 in 8000 α-particles get deflected through very large angles (over 90°), while all the others passed through with very little deflection. Rutherford was convinced that the majority of the mass was concentrated in a minute, positively charged region (the nucleus) surrounded by electrons. When and only when such an α-particle (which is positively charged) approaches rather close to that positively charged nucleus, will it get repelled, and strongly enough to rebound at high angles. From the small number of repelled α-particles, Rutherford deduced (a bound for) the size of the nucleus. The nucleus was found to be less than approximately 10^{-14} m in diameter.

The work had an immediate success. After all, the experiments had revealed how all matter is structured and thus affected every scientific and engineering discipline, making it one of the most pivotal scientific discoveries of all time. The astronomer Arthur Eddington called Rutherford's discovery "the most important scientific achievement since Democritus proposed the atom ages earlier".

A new paper by Geiger and Marsden, "The laws of deflexion of α-particles through large angles" in 1913, described a series of experiments which put the Rutherford model to the test Geiger and Marsden (1913). Now they could accurately measure the scattering pattern of the α-particles and they could count scintillations from every angle. They measured how the scattering pattern varied with the square of the nuclear charge, and in this way, tested whether the scattering was proportional to the atomic weight squared. They also found that the number of scintillations was proportional to the inverse of the fourth power of the speed, as predicted by the Rutherford scattering formula.

Rutherford's atomic model would eventually lead to more problems than it was able to solve. Some people started to think of the Rutherford atom as resembling the Solar System, with charged particles (electrons) orbiting the much bigger, heavier, slower nucleus (which would then correspond to the Sun). Such a Copernican view of the atom was not Rutherford's own conclusion, and yet it still lives on in popular imagination. More than once, we see advanced physics teachers explain how electrons orbit the nucleus, much like planets around the Sun (see Fig. 5.2). According to classical physics, at least at first sight, it was in fact impossible to imagine charged particles orbiting the atomic nucleus, as they would then necessarily radiate electromagnetic waves: an orbiting electron would thus lose energy and spiral into the nucleus. To fix this problem, scientists had to incorporate quantum mechanics into Rutherford's model. Clearly, chemistry was about to receive a whole new foundation.

7.2.2 Standard Model Experiments

Low–temperature physics was an important ingredient in establishing the standard model of elementary particles. Famously, the experiment by Chien-Shiung Wu (1912–1997) established that conservation of parity was violated (P-violation) by the weak interaction. The experiment was carried out in 1956 at the National Bureau of Standards low-temperature laboratory in Washington DC. It was experimental confirmation of the prediction by Tsung-Dao Lee and Chen-Ning Yang, who contributed significantly to statistical mechanics (Lee–Yang theorem, Yang–Baxter equations, etc.) and to subatomic particle physics (Yang–Mills theory) more generally. Lee and Yang had the idea of parity nonconservation and they actually proposed the experiment. As a result, they received the 1957 Nobel Prize in Physics. Chien-Shiung Wu, who designed and performed the experiment, was awarded the first Wolf Prize.[3] On the road to the Standard Model, it is said that Wu's discovery was the most significant since the Michelson–Morley experiment.

Particle scattering developed into Big Science projects with nuclear physics and elementary particle physics dominating the science budgets. It was claimed that fundamental physics was synonymous with high-energy and subatomic physics, and the

---

[3] Wu was called the Chinese Marie Curie, and Lise Meitner (1878–1968), another great experimentalist in nuclear physics, was called the German Marie Curie. Yet, in contrast with Marie Curie (double Nobel award winner), neither Wu nor Meitner were rewarded with the Nobel prize. Meitner collaborated with Otto Hahn on nuclear fission, and Hahn did receive the Nobel prize in 1944.

military powers of the day agreed. Around 1970, new accelerators were built. In 1972, the proton accelerator at Brookhaven was able to deliver beams of unprecedented intensity. In Europe, the proton accelerator at CERN produced intense beams, and is still a key component in CERN's accelerator complex. There was also the high-energy proton accelerator at Fermilab, and SPEAR, an electron–positron collider at the Stanford Linear Accelerator Center.

All that went together with experimental discoveries, clues for and confirmations of the Standard Model of elementary particle physics. Good news was the discovery in CERN in 2012, using the Large Hadron Collider, of the Brout–Englert–Higgs particle, predicted to exist and play a fundamental role in the Standard Model some 50 years earlier.

7.3   Nuclear Chain Reactions

On December 2, 1942, the world's first self-sustaining, controlled nuclear chain reaction occurred, paving the way for a variety of advancements in nuclear science and modern war crimes. The experiment managed by some 50 scientists, mostly involved in monitoring the reaction, took place in a converted squash court at the University of Chicago's abandoned Stagg Field.[4] It went under the direction of Enrico Fermi. The area was already densely populated in those days, but it seems that everyone concerned, and in particular Compton, professor at Chicago, trusted the calculations made by Fermi and Szilard. There was of course also a sense of urgency. Compton later explained:

> As a responsible officer of the University of Chicago, according to every rule of organizational protocol, I should have taken the matter to my superior. But this would have been unfair. President Hutchins was in no position to make an independent judgment of the hazards involved. Based on considerations of the University's welfare, the only answer he could have given would have been — no. And this answer would have been wrong.

Chicago Pile-1 was the world's first nuclear reactor to go critical. The word "pile" (heap, as Fermi called it) stands for "reactor", the main device that controls the nuclear fission reaction. The absorption of a neutron by a uranium-235 nucleus turns that nucleus into an excited uranium-236 nucleus. Uranium-236 rapidly splits into fast-moving lighter elements (fission products) and releases several free neutrons, and a large amount of electromagnetic energy. The chain reaction occurs because

---

[4] That was the university's football field, but by 1942, the football competitions had been discontinued.

fissioning uranium atoms emit neutrons. There is a burst of energy and the reaction continues because the average number of neutrons emitted per fission of the uranium-235 nucleus is known to be about 2.4. Criticality is achieved when the rate of neutron production is equal to the rate of neutron loss, including both neutron absorption and neutron leakage.

When he was working in Rome, Fermi discovered that slowing down neutrons via collisions makes them more likely to be captured by uranium nuclei, causing the uranium to fission.[5]

In general, a neutron moderator reduces the speed of fast neutrons, turning them into thermal neutrons, which are, however, far more likely than fast neutrons to propagate a nuclear chain reaction of uranium-235. They went for graphite as moderator[6] for an interesting and important reason. The reactor contained 45 000 ultra-pure graphite blocks. The use of graphite as a moderator was a key innovation and a great success. It was Leo Szilard (1898–1964) who had suggested to Fermi that they use carbon in the form of graphite as a moderator. Szilard had heard that National Carbon, a chemical company, would be able to produce graphite of sufficient purity. This was thanks to a man called Herbert MacPherson (1911–1993), hired as a physicist, which was and maybe still is unusual, to research carbon arc lamps, a major commercial use for graphite at that time. Because of his work studying the spectroscopy of the carbon arc,[7] MacPherson knew that the main relevant contaminant was boron, both because of its concentration and its affinity for absorbing neutrons, confirming one of Szilard's suspicions. Using purified graphite as a neutron moderator gave them an important lead over the Nazi nuclear program.[8]

The pile ran for about 4.5 min at about 0.5 W. Unlike most subsequent nuclear reactors, it had no radiation shielding or cooling system, as it operated at such very low power. By telephone, Compton notified James B. Conant (1893–1978), who had

---

[5] The history of nuclear fission involves more than just its discovery in 1938 by Otto Hahn, Fritz Strassmann, Lise Meitner, and Otto Robert Frisch. Enrico Fermi and his colleagues in Rome (the Via Panisperna boys, movie from 1989 by Gianni Amelio) bombarded uranium with neutrons, and Fermi concluded that the experiments had created transuranium elements ausenium (now known as neptunium) and hesperium (now known as plutonium). That was also the reason for his 1938 Nobel Prize in Physics, for "demonstrations of the existence of new radioactive elements produced by neutron irradiation, and for his related discovery of nuclear reactions brought about by slow neutrons." It turned out, however, that the new elements that Fermi had found were fission products.

[6] We also know of graphite-moderated reactors from major nuclear disasters: an untested graphite annealing process contributed to the Windscale fire in the United Kingdom in 1957, where the uranium fuel caught fire, and a graphite fire during the Chernobyl disaster of 1986 contributed to the spread of radioactive material.

[7] The carbon arc light was invented by the poet and chemist Humphry Davy around 1809 and consists of an arc between carbon electrodes in air. It was the first practical electric light, starting around 1870.

[8] The crucial information concerning boron impurities in graphite and how to get rid of them was not known to the German scientists. MacPherson knew a lot about carbon arcs as they were used in the movie industry and he had been asked to study their spectra.

worked on poisonous gases for the U.S. military in 1917–18. President of Harvard University and chairman since 1941 of the National Defense Research Committee, Conant oversaw vital wartime research projects. The secret development of the reactor was the first major technical achievement for the Manhattan Project. The conversation was in an impromptu code:

Compton: The Italian navigator has landed in the New World.
Conant: How were the natives?
Compton: Very friendly.

Chapter 8
Light in the Universe and the Invariance of Proper Time

The unification of electricity, magnetism, and optics (which we may abbreviate to EMO; see Chap. 3) brought new problems into the foreground. The most important of these was the invariance of "the speed of light," or rather, the fact that Maxwell's equations do not satisfy the Galilean rule of velocity addition. Put another way, EMO is not invariant under Galilean transformations. Furthermore, from many examples, both thought experiments and real observations, it was clear that electric and magnetic fields were in some sense interchangeable, or at least involved some kind of symmetry. It was tempting to think then that further unification was possible, in which the individual identities of the electric and magnetic fields would disappear and make room for a unique electromagnetic field with some specific transformation properties when viewed from different observer frameworks.

A better understanding of these issues gave rise to the special theory of relativity. Its history is closely connected with electrodynamics and the properties of light. That is the background and context of the present chapter. Among the foundational explorations, there was the crucial experiment by Michelson and Morley:

> The velocity of light is one of the most important of the fundamental constants of Nature. Its measurement by Foucault and Fizeau gave as the result a speed greater in air than in water, thus deciding in favor of the undulatory and against the corpuscular theory. Again, the comparison of the electrostatic and the electromagnetic units gives as an experimental result a value remarkably close to the velocity of light–a result which justified Maxwell in concluding that light is the propagation of an electromagnetic disturbance. Finally, the principle of relativity gives the velocity of light a still greater importance, since one of its fundamental postulates is the constancy of this velocity under all possible conditions.
> — A.A. Michelson, Studies in Optics.

In 1884, Lord Kelvin (William Thomson) was giving a master class on "Molecular Dynamics and the Wave Theory of Light" at Johns Hopkins University. Just as Kelvin was describing sound as waves of pressure in air, there was also an attempt to describe an electromagnetic wave equation for light. It was natural then to assume

© The Author(s), under exclusive license to Springer Nature Switzerland AG 2023
C. Maes, *Facts of Matter and Light*,
https://doi.org/10.1007/978-3-031-33334-7_8

a luminiferous æther susceptible to vibration. The study group included Albert A. Michelson (1852–1931) and Edward W. Morley (1838–1923). They would go on to do the experiment that would undermine that same æther theory.

8.1 Michelson–Morley Experiment

One of the recurrent questions in studies of our universe concerns the relationship between space and matter. At least since Newton, (real and thought) experiments have been set up to try to discover absolute motion. That came up again and with more force when reflecting on the medium in which light might propagate. According to Maxwell's theory, light is a wave, similar to the ripples on the surface of a lake making a transverse wave. Similarly, sound waves propagate as longitudinal pressure waves through air or through some other material. The medium that was supposed to support transverse electromagnetic waves was called æther. The æther was to permeate all space and be the stationary carrier of light waves. That was the prevalent æther view at the time. The specific issue then, toward the end of the nineteenth century, was to discover motion with respect to that supposed æther.

The basic technology and innovation that was necessary for the Michelson–Morley experiment was interferometry. Here, Michelson was a pioneer Michelson (1881). In the Michelson interferometer, an interference pattern is obtained by dividing a photon beam into two distinct beams, directed along different paths, after which they converge again. See Fig. 8.1. The two paths are such that their lengths may be different or they may go via different materials. That would produce a phase shift in the optical path, and hence a pattern.

That invention allowing such great precision is so clever that it has been used in many other experiments right up to this day to make the more precise "discrimination measurements." For example, the Michelson interferometer scheme is used to detect the passage of gravitational waves in LIGO and VIRGO. The passage of a gravitational wave causes a tiny deformation of space, and hence a very small difference in the optical paths of the laser beams in the two arms of the interferometer. This difference is measured by optical sensors via changes in the interference fringes.

The experiment was performed in 1887 by Michelson and Morley in Cleveland, Ohio, at what is now the Case Western Reserve University. It has been called the most famous failed experiment. It set out to determine the existence of the light-bearing (luminiferous) æther; it failed. Some experiments show that certain things exist, others that a certain thing does not exist.

The idea was to measure the speed of light in two perpendicular directions. If there was a difference, it would show the relative motion of matter (the Earth in particular) through the æther. It would catch more "æther wind" one way than the other (see Fig. 8.2). The speed of light in the direction of motion through the presumed æther and the speed at right angles would differ. Michelson and Morley found no significant difference: a null result Michelson and Morley (1886, 1887) (see Fig. 8.3).

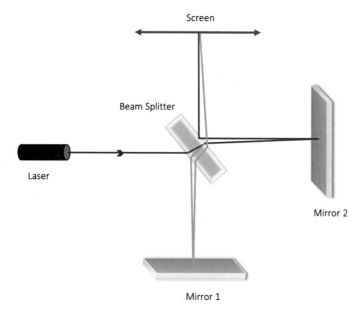

Fig. 8.1 Michelson interferometer

With hindsight, the Michelson–Morley result is often seen as strong to convincing evidence against the æther theory. It paved the way for the relativity revolution, making it easier to swallow, so to speak. While Einstein probably had not heard about the Michelson–Morley experiment in 1905, when he wrote down the basic ideas of special relativity, he would later write:

> If the Michelson–Morley experiment had not brought us into serious embarrassment, no one would have regarded the relativity theory as a (halfway) redemption.

Then, something remarkable happened, although it was not particularly unusual. Typically, the physicist's best instinct is to stay with the old theory and try to save as much as possible. Neither Michelson nor Morley ever believed, or even considered, the possibility that their results implied that the "luminiferous æther" did not exist. They were not alone; many excellent classical physicists started to introduce new effects to save the æther. For example, in 1889, George Francis FitzGerald (1851–1901) postulated a contraction of objects in the direction of their motion in inertial frames of reference. Hendrik Lorentz (1853–1928) postulated something similar, seeking a way out through what were at the time the latest insights into the electrical structure of matter. To obtain the null results for the Michelson–Morley experiment, Lorentz concluded that an object moving in the same direction as the Earth (opposite the æther) would shrink by some factor. Both Lorentz and FitzGerald assumed that the contraction was the result of changes in the molecular forces.

Over the following years, Lorentz developed the theory and first in 1899 and then in 1904 he deduced the length contraction from more general transformation

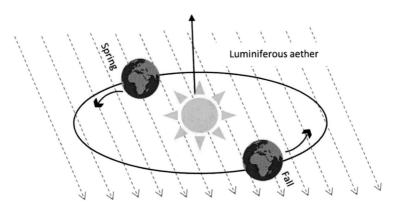

Fig. 8.2 A depiction of the concept of the "æther wind"

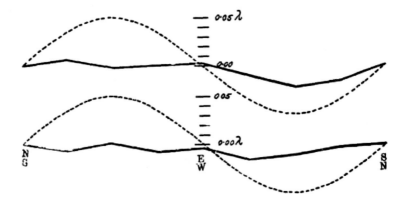

Fig. 8.3 From "On the relative motion of the Earth and the luminiferous æther," summarizing Michelson and Morley's results. Their observations at noon give the upper solid line, and their evening observations are summarized in the lower solid line. It is important to note that the theoretical curves and the observed curves are not plotted on the same scale: the dotted curves represent only one-eighth of the theoretical displacements. Source Michelson and Morley (1887)

formulas, now known as the Lorentz transformations. They gave the coordinates of one object that moves with respect to the æther in terms of the same at rest relative to the æther. Those formulas had already appeared in work by Woldemar Voigt (1850–1919) in 1887 and Joseph Larmor (1857–1942) in 1900 and could perfectly explain the zero result in the Michelson–Morley experiment. Lorentz, however, continued to hold a dynamic view of what was going on. There was assumed to be a physical cause for those transformations, something between the æther and the electrons of the moving object. The æther still functioned as an absolute frame of reference, and he stuck to absolute simultaneity.

As with all famous experiments, the Michelson–Morley experiments have also been repeated many times, with steadily increasing sensitivity. Together with some other experiments (such as the Ives–Stilwell and Kennedy–Thorndike experiments of 1932), the Michelson–Morley experiments constitute one of the fundamental tests of the æther hypothesis. From a modern perspective, what the Michelson–Morley experiment showed can be referred to as the absoluteness of a distance, the relativistic invariant interval. This is what is explained below.

8.2 Special Relativity

> For the rest of my life I will think about what light is.
> — A. Einstein, as quoted by W. Pauli, *Aufsätze under Vorträge über Physik und Erkenntnis-theorie* (1961), p 88.

Special relativity grew out of classical mechanics from experimental facts concerning electrodynamics and light phenomena. Various physicists were involved, including Hendrik Lorentz and Henri Poincaré, but the main architect was Albert Einstein for his daring and realistic attitude toward the meaning of time:

> The special theory of relativity meant the dethronement of time as a rigid tyrant imposed on us from outside, a liberation from the unbreakable rule of "before and after." For indeed time is our most severe master by ostensibly restricting the existence of each of us to narrow limits — 70 or 80 years, as the Pentateuch has it. To be allowed to play about with such a master's program believed unassailable till then, to play about with it albeit in a small way, seems to be a great relief, it seems to encourage the thought that the whole "timetable" is probably not quite as serious as it appears at first sight. And this thought is a religious thought, nay I should call *the* religious thought.
> — E. Schrödinger, Tarner Lecture 1956.

After 1905, Paul Langevin, Max Planck, and Wolfgang Pauli were among the most active forces spreading the word. However, apart from Einstein's 1905 paper, *On the electrodynamics of moving bodies*, the next most important addition to the theory was the introduction of special relativistic spacetime by Hermann Minkowski (1864–1909). His lecture *Raum und Zeit* began with the celebrated words:

> The views of space and time which I wish to lay before you have sprung from the soil of experimental physics, and therein lies their strength. They are radical. Henceforth space by itself, and time by itself, are doomed to fade away into mere shadows, and only a kind of union of the two will preserve an independent reality.
> — Hermann Minkowski, Köln, 1908.

Note in particular that the geometer Minkowski stresses the experimental points. Special relativity is indeed best formulated as being directly linked to observation and measurement. The rest of this chapter looks quite theoretical, but note how almost all the ideas are directly linked to operations of measurement or observation. That is the reason for including all this.

8.2.1 *Popular Relativity*

Popular accounts of special relativity center around a number of slogans such as nothing can go faster than light, the speed of light is absolute and the same for all observers, moving clocks run slow, moving rods are shorter, energy is mass, and so on, which are at best imprecise. However, they do serve as a good introduction to certain phenomena that constitute the empirical content of the special theory of relativity. Basically, they can all be "derived" by modifying standard Newtonian mechanics by the sole introduction of the light postulate, that the speed of light is independent of the state of the source. Let us illustrate that with a couple of examples.

How do we measure time? We need clocks, and a good reliable clock is a light clock, as it turns out to be so independent from the state of motion. Imagine a rocket moving at speed v in a horizontal direction away from the Earth and let d be a distance in the vertical direction, for example, between floor and ceiling mirrors inside the rocket. Let us denote by t the time observed on Earth for a light pulse to go that distance d, while t' is the time measured by an astronaut in the rocket. As the rocket is moving at speed v, we have

$$(vt)^2 + d^2 = (ct)^2 \, ,$$

by Pythagoras, where c is the speed of light, unchanged from the point of view of the Earth. On the other hand, $d = ct'$, from which we conclude the famous formula for time dilation:

$$t' = t\sqrt{1 - \frac{v^2}{c^2}} \, ,$$

in a rocket moving at speed v away from the Earth. As $t' < t$, we say that moving clocks run slow, i.e., the Earth-bound observer sees a greater time interval for the light to move between the ceiling and floor in the rocket. Of course, the effect depends on the ratio v/c, and is too small to notice in everyday circumstances (remember that $c = 300\,000$ km/s). However, this time dilation is observed in various precision measurements, for example, for the decay time of particles resulting from cosmic radiation (such as the muon).

In the above thought experiment, we assumed that the distance d remained the same for both observers. That is because the motion is orthogonal to that distance. However, to measure distances in the horizontal direction, we can again use light pulses sent along rigid rods. But as the time runs more slowly for moving clocks, we must necessarily find that lengths contract. That is the Lorentz–FitzGerald contraction.

8.2.2 Minkowski Spacetime

There is a huge body of literature explaining special relativity. No need to expand on this, on the contrary. For those who seek a more conceptual line of presentation, the pleasant book by Tim Maudlin, Maudlin (2015), also inspired the very brief exposition below.

We restrict ourselves to the minimal elements and start with a fact or physical phenomenon that can be checked experimentally:

The trajectory of light in a vacuum is independent of the physical state of its source.

In other words, the light trajectory does not depend on what the source was doing when the light was emitted. The color (wavelength) can change depending on whether the source comes towards us or recedes from us, but the trajectories cannot change; in particular two pulses of light arrive at the observer together even if they were emitted, say, from two flashbulbs passing each other in opposite directions.

As it seems natural to say that, in vacuum, there is no physical structure except for the structure of spacetime itself, it follows that it is the geometry of spacetime that determines the trajectory of light rays.

Let us then characterize spacetime by light properties. More precisely, with any spacetime point p, we associate the future light cone, i.e., the set of points where light can go when emitted from p in any possible direction. Similarly, there is a past light cone at p containing all light trajectories that can reach p. That is the first structure of our new spacetime: there is a past and future light cone associated with each event, i.e., each spacetime point.

The spacetime of special relativity is Minkowski spacetime and we use Lorentz coordinates to assign numbers to events. The spacetime is 4-dimensional, so we need 4 coordinates, denoted by t, x, y, z. A point moves around continuously in Minkowski spacetime if and only if its Lorentz coordinates all change continuously as it moves. These are just the same as Cartesian coordinates for the Euclidean space E^4. The straight lines are also similar: a set of events in Minkowski spacetime forms a straight line if the coordinates of the events are of the form ($t = a_1 s + a_2, x = b_1 s + b_2, y = c_1 s + c_2, z = d_1 s + d_2$) with at least one of the $a_1, b_1, c_1,$ or d_1 nonzero. That means that the topological and affine structure of Minkowski spacetime is just the same as for Galilean spacetime. And now comes what is truly new: instead of the Euclidean distance we use the function

$$d(p,q) := \sqrt{c^2[t(p) - t(q)]^2 - [x(p) - x(q)]^2 - [y(p) - y(q)]^2 - [z(p) - z(q)]^2} ,$$

known as the invariant relativistic interval between the events p and q. The number c is called the speed of light, but is of no great importance yet. Observe that $d(p, q)$ is not a metric (a distance). The value $d(p, q)$ can be zero, for example, even between non-identical spacetime events. In fact, the equation $d(p, q) = 0$ for fixed p defines the light cones at p. As the parameter $c \uparrow +\infty$, the light cones open further and further to become equal t-slices in the limit.

We are ready to formulate the basic postulates:

Law of Light . The trajectory of a light ray emitted from an event (in a vacuum) is a straight line on the future light cone of that event.

The light cone at an event partitions the remainder of spacetime into five topologically connected sets: (1) the events on the future light cone, (2) the events within the future light cone, (3) the events on the past light cone, (4) the events within the past light cone, and (5) the events outside the light cones. There is some terminology here: a time-like separation is one between an event and any event within its future or past light cones; events are said to be light-separated when one is on the past or future light cone of the other, and space-like separated when one is outside the light cones of the other. Note that values of $d(p, q)$ vary accordingly between real positive, zero, and imaginary.

The fact that "nothing can go faster than light" can be expressed in terms of the geometry as follows:

Limiting Role of the Light Cone. The trajectory of any physical entity that goes through an event never goes outside the light cone of that event.

Similarly to Newtonian mechanics (see Sect. 2.3.2):

Relativistic Law of Inertia. The trajectory of any physical entity subject to no external influences is a straight line in Minkowski spacetime.

We now move to the connection between the geometry of spacetime and the observable behavior of material objects. We will come back to this later, but for now we formulate it as follows:

Clock Hypothesis. The amount of time that an accurate clock shows to have elapsed between two events is proportional to the interval along the clock's trajectory between those events; or, in short, clocks measure the interval along their trajectories.

More specifically, we take

$$\tau = \frac{d(p, q)}{c} .$$

This hypothesis makes sense from the fact that two clocks side-by-side measure the same elapsed "time"—well, we call it the proper or eigen time. Clocks that wander off along different trajectories through spacetime can have recorded quite different elapsed times if they ever reunite.

8.2.3 Twin Paradox

The twin paradox, originally due to Paul Langevin (1872–1946), is a thought experiment in special relativity involving identical twin brothers, one of whom makes a journey into space in a high-speed rocket and returns home to find that the twin who remained on Earth has aged more. The reason is simple; the length of the trajectory of the traveling twin is shorter. The twin remaining at home could also be accelerating, going back and forth between Earth and Moon for example, but as long as his spacetime trajectory has a longer *proper* time, he will be older than his traveling brother.

Let us take for example two trajectories that connect event $(0, 0)$ with event $(10, 0)$. (We forget about the y, z-coordinates.) If Peter moves along the $x = 0$ axis, as he likes to do (he's a stay-at-home), the proper time elapsed is

$$\sqrt{(10 - 0)^2} = 10 ,$$

where we have put $c = 1$ for concreteness. In the same way, we consider the trajectory of Paul, traveling in a straight line from $(0, 0)$ to $(5, 3)$ and then from $(5, 3)$ straight to $(10, 0)$. The proper time elapsed here is

$$\sqrt{(5^2 - 3^2/c^2)} + \sqrt{(10 - 5)^2 - 3^2/c^2} = 8 ,$$

It does indeed turn out that the proper time experienced by Paul (the traveler) is smaller. This has a relativistically invariant meaning, in that the same numbers are obtained after Lorentz transformations. In this sense, Paul and Peter are of course not symmetric or interchangeable, but their true difference is not to be found in words such as "Paul is travelling" or "Paul is accelerating," which would lead to symmetric treatments.

By the way, not everything is relative in relativity theory. There is for example something objective about acceleration in special relativity.

8.2.4 NonEuclidean Geometry

The Minkowski equivalent of a Euclidean measure for distance can be found in the invariant relativistic interval. We take a large collection of identically constructed alarm clocks, all set to go off one minute after a button is pushed. At a certain moment, the button is pushed and all clocks move in all directions without further forces. The shape in spacetime of the ringing events is a hyperboloid of revolution, for example characterized by the equation

$$c^2 t^2 - x^2 - y^2 - z^2 = 1 .$$

Note that, if the number c were very big, i.e., $c \uparrow +\infty$, that would amount to flattening the hyperboloid, and we would have the Galilean measure of time. However, for every finite value of c, we have a locus of events described by a hyperboloid of revolution. All points on that hyperboloid are all exactly at the same "distance" from the origin in the intrinsic geometry of Minkowski spacetime. That is the origin of the nonEuclidean geometry in special relativity.

8.3   And More Generally

Relativity grew from the confrontation between electromagnetism and mechanics. There is the special theory, with the introduction of Minkowski spacetime, which enters the formalism of quantum field theory. After all, high-energy physics requires relativistic calculations. Scattering and collision theory for highly energetic particles requires us to use the prescriptions of special relativity. This combination of special relativity and quantum mechanics, in the attempt to build a unifying framework between electromagnetic, weak, and strong interactions, has resulted in the Standard Model of elementary particle physics.

The second part of relativity is the general theory. It entered physics as the theory of spacetime, a theory of gravity.

Remember where gravity brought us in Sect. 2.4. The fact that the free motions (straight lines) are those of free fall tells us immediately that the structure of spacetime will be strange, or at least stranger than the one on the Euclidean blackboard. For example, despite being "straight," the worldline of a satellite orbiting the Earth repeats itself again and again, and two satellites orbiting in different directions move along straight lines that repeatedly intersect each other. Parallel straight lines that intersect each other are only possible in non-Euclidean geometries. That is why and how non-Euclidean geometry enters crucially into the description of the spacetime structure of general relativity.

Once again, the straightness of the satellite worldlines follow from the decision to eliminate any force of gravity. These free-fall trajectories are the equivalent of straight time-like trajectories in special relativity. Similarly, free-fall trajectories are locally maximal: any small deviation will shorten their length. Thus, a clock on such a trajectory will run off more time than nearby accelerated clocks. Here is an example, due to Richard Feynman (1995). Consider a clock "at rest" in your left hand and another clock in your right hand. The latter is thrown up at exactly 12h00 and lands back in your hand when the clock in your left hand shows exactly 12h01. The thrown clock will show more elapsed time.

General relativity also provides the classical framework for the standard model of physical cosmology. It is not true that the geometry of spacetime is *determined* by the distribution of matter, but it is influenced by it. The Einstein field equation states (only) that the average amount of curvature in a spacetime region is determined by the amount of matter and energy in that region. The more matter and energy there is, the more curvature there is. In fact, matter and energy tend to *focus* locally-parallel inertial trajectories. That is what we see as the universal attraction between bodies, called gravity.

In Minkowski spacetime, the curvature is uniformly zero everywhere. It is not known whether or how general relativity breaks down for very high curvatures or near singularities in the spacetime structure. That is why new physics is needed and is thought by many to be discoverable in studies of black holes. The main characteristic of a black hole being that everything falls freely into it.

General relativity has shaped much of our understanding of the universe, especially its dynamics. It shows that the universe has been expanding from an earlier hot and dense universe. But beyond that, much remains unclear. There is certainly not a shred of empirical evidence for the big bang:

> The first thing to understand about my field is that its name, Big Bang theory, is quite inappropriate. It's very unfortunate that one thinks of the beginning, whereas in fact, we have no good theory of such a thing as the beginning.
> — James Peebles, Nobel Prize 2019.

The idea of a beginning may be attractive but has no scientific support—there is no proof of an origin event. That must be distinguished from the evidence for an expanding universe, which was hot and dense some 14 billion years ago, i.e., there is abundant empirical evidence for a big bang theory.

> God's daughter in her pinafore kept blocks,
> With which up in the clouds she had learned to play.
> But when, tired and bored, she put them away
> She couldn't fit all of them into the box
>
> In a proper, neatly ordered display.
> Now God was asleep, strict and orthodox.
> So feeling safe, she dropped them, sly young fox,
> Made straight for a fine angel made of clay.
>
> The blocks then tumbled through the cosmic void,
> Arriving at an empty planet, where
> They stayed right in position as they'd been hurled.
>
> Most fragments as hills and dales were deployed;
> The bits that were intact formed here and there
> Great cities and small hamlets through the world.

— Creation Myth, by J.J. Slauerhoff (1898–1936), Translation by Paul Vincent, 2007.

Observations and confirmations of the standard cosmological model abound since the 1960s, while important puzzles, such as the baryon (matter-antimatter) asymmetry problem, the origin of the space roar (residual background of radio synchrotron radiation) and the (changing and accelerating) rate of cosmic expansion, cast doubts on perhaps oversimplified approaches that ignore inhomogeneities or fluctuations or nonequilibrium effects dating from the early universe. In the coming decades we may expect more news from that early universe, perhaps even from plasma and quantum optics experiments or from numerical work simulating (even only partially) relevant features.

Still the question comes up again, does the æther exist? It cannot be molecular or ponderable, but can it not be the word for vacuum. It appears as the passive medium for propagation of light and gravitational waves. Yet, by quantum effects, is then that æther noisy and active?

Chapter 9
Dynamical Activity of the Vacuum

The idea that the vacuum does not exist or is not so terribly empty is rather old. Since time immemorial, there has existed a feeling of *Horror Vacui*, in the arts[1] as well as in decoration more generally; for example in Fig. 9.1.

In physics, Nature's abhorrence of the void has been discussed since time immemorial, while the presence and necessity of *empty space* have been the subject of both theoretical research and a long series of pneumatic and barometric experiments. One important instance was the experiment of Evangelista Torricelli (1608–1647) in 1643, demonstrating the existence of the vacuum. Or again, recall the expedition by Blaise Pascal (1623–1662) on the Puy-de-Dôme in 1648, showing that atmospheric pressure lessens with altitude.[2]

It is one thing to accept the existence of the vacuum, and it is a quite different matter to describe its properties or its role, even when there is convincing experimental evidence. Absence of air is obviously not the same as empty space. Regarding a more modern debate, think about the idea of the æther in Sect. 8.1, which was supposed to be an all-pervading fluid. For Lorentz , for example, the æther was "the seat of an electromagnetic field with its energy and its vibrations".

Another example is today's debate about dark energy, whether it exists as an intrinsic large scale and extensive property of the vacuum. The question is related to the accelerated expansion of the universe; it is perhaps even another word for it. Surely, many would assert that dark energy is not a quantum vacuum energy in the sense that will follow below.

Toward a more modern understanding, the concept of a zero-point energy first appeared in Planck's "second" quantum theory of 1911. Toward the second half of

---

[1] As a visit to the basilica of San Juan de Dios in Granada makes very clear.

[2] The book *Expériences nouvelles touchant le vide* (1647) provided reasons to believe in a vacuum. His health did not allow Pascal to climb the Puy–de-Dôme, but his brother-in-law was found able to carry out the mission.

© The Author(s), under exclusive license to Springer Nature Switzerland AG 2023
C. Maes, *Facts of Matter and Light*,
https://doi.org/10.1007/978-3-031-33334-7_9

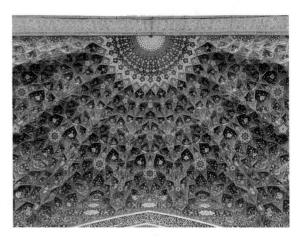

the twentieth century, physicists began to speak about empty space as characterized
by vacuum fluctuations, a perspective offered by quantum field theory, especially
within its framework of particle creation and annihilation.

9.1 Beginnings

The world does not come to a halt at absolute zero. Particles and fields remain
active in their states of lowest energy. Vacuum fluctuations in general refer to the
fact that the expected kinetic energy in quantum ground states does not need to be
zero. Observables maintain a nonzero variance when statistically evaluated using
the appropriate wave function, even for the ground state (lowest energy). *Forever
dancing* ..., condemned as the girl with the red shoes in Andersen's terrifying and
infuriating fairy tale.

The idea of zero-point[3] fluctuations was born in the quantum writings of Max
Planck (1911–13).[4] We refer to Norton (1993), Kragh (2023) for references and
more details. The hypothesis originates in the idea of a discontinuous emission (and
not absorption) of radiation, which would only occur when the energy of an oscillator is an integer multiple of $h\nu$ (constant energy difference). The values of the
quantized energy levels of the harmonic oscillator exhibit a zero-point value. Here
is the punchline:

---

[3] Note, however, that it was only from approximately 1925 that it became commonplace to refer to
"zero-point energy.".

[4] Earlier in Sect. 5.2, we referred to the complicated history of blackbody radiation. For Planck,
part of the motivation certainly concerned the foundations of thermodynamics, and it would seem
he was ready to call the zero-point energy an "indispensable element.".

This rest energy remains with the oscillator, on the average, at the absolute zero of temperature. [The oscillator] cannot lose it, for it does not emit energy so long that the energy is smaller than $h\nu$.

It is like having all sorts of coins in your wallet, while being allowed to pay only with 1 euro coins (or some other unit value); there will always remain some small change. Planck admitted, however, that it had no experimental basis. Still, in a letter to Heike Kamerlingh Onnes in Leiden, in March 1915, Planck wrote:

I have almost completed an improved formulation of the quantum hypothesis applied to thermal radiation. I am more convinced than ever that zero-point energy is an indispensable element. Indeed, I believe I have the strongest evidence for it.

In 1913, a paper by Einstein and Otto Stern set out to calculate the specific heat of hydrogen using zero-point fluctuations. Einstein soon had second thoughts, writing to Paul Ehrenfest that he considered the idea "as dead as a doornail."

So it looked at first as though Planck's idea led to a dead end. But then, there appeared a first shift in the phrasing. In 1920, Einstein declared that general relativity needed to include the zero-point energy of empty space:

According to the general theory of relativity, space without æther is unthinkable; for in such space there not only would be no propagation of light , but also no possibility of existence for standards of space and time (measuring rods and clocks), nor therefore any spacetime intervals in the physical sense.
— A. Einstein (1920), Äther und relativitäts-theorie.

It is useful therefore to distinguish the zero-point energy of matter (Planck's oscillators) and the zero-point energy of space. The latter was an idea[5] that went back to Walther Nernst , who had written about a cosmic æther filled with a huge amount of zero-point energy to save the universe from an inevitable heat death, as originally speculated upon by Kelvin , Boltzmann, and others:

The atoms of all elements of the universe will sooner or later entirely dissolve in some primary substance [Ursubstanz], which we would have to identify with the hypothetical medium of the so-called luminiferous æther. In this medium [...] all possible configurations can presumably occur, even the most improbable ones, and consequently, an atom of some element (most likely one with high atomic weight) would have to be recreated from time to time. [...] This means, at any rate, that the cessation of all events no longer needs to follow unconditionally as a consequence of our present view of nature.
— W. Nernst , 1912. Quoted in Kragh (2023).

Pleasantly, the zero-point energy in space constantly interacts with matter, rather as an "infinite and active reservoir":

Our eyes need not, in the far future, have to look at the world as a horrible graveyard, but as a continual abundance of brightly shining stars which come into existence and disappear.
W. Nernst , 1916. Quoted in Kragh (2023).

---

[5] Note, however, that Nernst never spoke about general relativity, nor ever mentioned the cosmological constant.

That brought about a new interpretation of vacuum energy, as a "sea" and reservoir of (quantum) matter. In 1927, Paul Dirac constructed the mathematical and conceptual basis of zero-point fluctuations, filling the physical vacuum with virtual particles and antiparticles. It was the birth of quantum electrodynamics, the theory that would give a meaning to the dynamical activity of the vacuum, as realized, for example, in the phenomenon of vacuum polarization.

9.2 Lamb Shift

Energy levels of the hydrogen atom and spectra more generally are measured in interaction and exchange with electromagnetic fields. In Dirac 's theory of atomic hydrogen, the levels $2S\frac{1}{2}$ and $2P\frac{1}{2}$ have the same energy; see (5.2) with $n = 2$, $j = 1/2$ while $\ell = 0$ (for S) and $\ell = 1$ (for P) (see Fig. 9.2).

It was believed from measurements in the 1930s and 1940s that the $2S\frac{1}{2}$ and $2P\frac{1}{2}$ levels in the hydrogen atom were slightly shifted. There had already been attempts to calculate radiative corrections in the 1930s. In 1938, Robley Williams went down to temperatures of 100 K to avoid Doppler broadening while measuring the Balmer spectrum (5.1) from a deuterium discharge lamp Williams (1938) (see Sect. 5.1). He was able to see the $3P\frac{1}{2} \rightarrow 2S\frac{1}{2}$ transition for both deuterium and hydrogen. Interestingly, the Dirac theory predicts that the separation between the $3D\frac{3}{2} \rightarrow 2P\frac{1}{2}$ transition and the $3P\frac{1}{2} \rightarrow 2S\frac{1}{2}$ transition just equals 3.24 GHz, which is the energy difference between the $3D\frac{3}{2}$ and the $3P\frac{1}{2}$ states. But as Williams remarked for his experiment:

> The most striking aspect of the data [...] is the large percent deviation from theory in the interval from component 2 to component 3.
> — R. Williams (1938) in Williams (1938).

That discrepancy seemed to indicate that the $2S\frac{1}{2}$ and the $2P\frac{1}{2}$ states were not degenerate (contradicting Dirac 's theory). That was the specific context for Lamb and Retherford (1912–1981) in 1947 to perform the experiment in which they directly measured the energy difference between the $2S\frac{1}{2}$ and $2P\frac{1}{2}$ states.

9.2.1 Lamb–Retherford Experiment

As one of the implications of war research efforts (radars!), studies in light–matter interactions were refined by orders of magnitudes, thanks to experimental breakthroughs made possible by advances in microwave technology.

In the case of Willis Eugene Lamb Jr (1913–2008), various competences and interests came together. He had already been interested in deviations from the Coulomb law in his thesis research. He knew about atomic spectroscopy and he knew about microwave engineering from his radar work at Columbia University. He joined

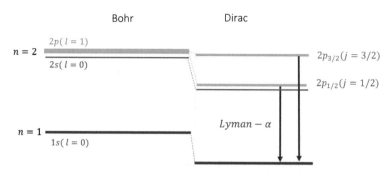

Fig. 9.2 Fine structure of energy levels in hydrogen. Relativistic (Dirac) corrections to the Bohr model : the Lyman–α line becomes a doublet

Columbia in 1938, and from 1943 to 1951, he worked with the Columbia Radiation Laboratory. His research was defense-related work on the radar. He was aware of the possible existence of a 2S–2P level shift, and he also understood that it was important:

> The hydrogen atom is the simplest one in existence, and the only one for which essentially exact theoretical calculations can be made on the basis of the fairly well confirmed Coulomb law of interaction and the Dirac equation for an electron. Such refinements as the motion of the proton and magnetic interaction with the spin of the proton are taken into account in rather good approximate fashion. Nevertheless, the experimental situation at present is such that the observed spectrum of the hydrogen atom does not provide a very critical test either of the theory or of Coulomb's law of interaction between two point changes. A critical test would be obtained from a measurement of the fine structure of the $n = 2$ quantum state,
> — W. Lamb , summer 1946 progress report.

In their experiment, Lamb and Retherford prepared a beam of hydrogen atoms in the $1S\frac{1}{2}$ ground state. When the beam was bombarded with electrons, some of the atoms were excited to the $2S\frac{1}{2}$ level, which has a lifetime of about 0.1 s. The atoms were then passed through a region of electromagnetic radiation with energy equal to the energy difference between the $2S\frac{1}{2}$ and $2P\frac{1}{2}$ states. This induced transitions between these two states. However, once an atom is in the $2P_{1/2}$ state, it decays to the ground state with a lifetime of 10^{-9} s. Finally, the atoms were made to strike a piece of tungsten foil and the atoms that were still in the $2S\frac{1}{2}$ state decayed to the ground state. They thereby liberated electrons from the foil in a process called Auger emission.[6] Comparing the emission current from the foil between the case with the electromagnetic radiation turned on (fewer electrons liberated) and with the radiation turned off (more electrons liberated), Lamb and Retherford determined the energy difference between the $2S\frac{1}{2}$ and $2P\frac{1}{2}$ states, with initial measurements indicating that it was about 1000 MHz MHz (see Fig. 9.3 and see also Hänsch 1979 for more details).

---

[6] The Auger electron is emitted when some of the energy released when a core atomic vacancy is filled with a higher energy electron is not totally transferred to a photon.

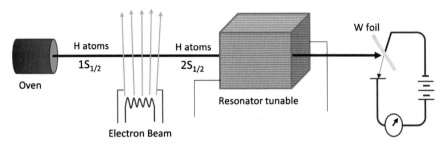

Fig. 9.3 The setup in the Lamb–Retherford experiment of 1947

Lamb won the Nobel Prize in Physics in 1955 "for his discoveries concerning the fine structure of the hydrogen spectrum." Lamb and Retherford had used high-precision microscopy to measure the surprising shift of electron energies in the hydrogen atom Lamb and Retherford (1947), finding a difference of about 1000 MHz MHz between the $2S\frac{1}{2}$ and $2P\frac{1}{2}$ hydrogen levels. This cried out for an explanation:

> It does not often happen that experimental discoveries exert an influence on physics as strong and invigorating as did your work . Your discoveries led to a re-evaluation and a reshaping of the theory of the interaction of electrons and electromagnetic radiation, thus initiating a development of utmost importance to many of the basic concepts of physics, a development the end of which is not yet in sight.
> — I. Waller , presentation speech Nobel prize to W. Lamb .

Not long after the experimental results were announced in 1947, Hans Bethe (1906–2005) published a nonrelativistic calculation in good qualitative agreement with the experiment Bethe (1947). It was inspired by Kramers , who showed how to subtract off infinities, an art that was in fact initiated by Jordan, in the context of the zero-point energy. This work was extended by various others, including Lamb himself, to a fully relativistic theory in quantitative agreement with experiment. Today, we call that quantum electrodynamics:

> Those years, when the Lamb shift was the central theme of physics, were golden years for all the physicists of my generation. You were the first to see that this tiny shift, so elusive and hard to measure, would clarify our thinking about particles and fields.
> — Freeman Dyson writing on Lamb 's 65th birthday.

We must add that the anomalous g-factors (magnetic moments for the electron or muon) are also due to radiative corrections, similar to the shift in the above hydrogen-atom levels. The Dirac equation predicts a magnetic moment equal to $g = 2$. Yet, the observed value differs, and is thus called the anomalous magnetic moment:

$$a = (g - 2)/2 = 0.001\,159\,652\,181\,643(764)\ ,$$

this being compatible with the calculations.

Lamb also contributed in many good ways to the theory of laser physics and quantum optics (fields of activity that did not yet exist in those days.). For example, he anticipated the discovery of the maser and the laser. A student of Lamb at Stanford, Theodore Maiman (1927–2007) at Hughes Research Laboratories, was the first to actually build a laser. Together with his employer, Hughes Aircraft Company, Maiman announced the laser to the world in a press conference in 1960.

9.2.2 Calculation

Let us add a sketch of the calculation. The heuristic argument goes as follows. The electron in the hydrogen atom is subject to the Coulomb potential $V(r) = -e^2/r$. Here, r can be viewed as the distance to the nucleus, but there will be fluctuations in that distance. If Δr is the displacement due to the fluctuations from the standard orbit, then the change in potential energy is

$$\langle \Delta V \rangle = \frac{1}{6} \nabla^2 V \langle (\Delta r)^2 \rangle ,$$

by Taylor expansion to second order. The energy shift is thus

$$\Delta \, \text{energy} = \frac{1}{6} \langle n\ell m_\ell | \nabla^2 V | n\ell m_\ell \rangle \langle (\Delta r)^2 \rangle . \tag{9.1}$$

But

$$\langle n\ell m_\ell | \nabla^2 V | n\ell m_\ell \rangle = 4\pi e^2 |\psi_{n\ell m_\ell}(r = 0)|^2 .$$

And here comes the main idea: only the wave functions for the S-states do not vanish at the origin, and there $|\psi_{n00}(r = 0)|^2 = 1/\pi n^3 a_0^3$, where a_0 is the Bohr radius. This already explains the shift, but we can get more quantitative.

We can estimate $\langle (\Delta r)^2 \rangle$ for each frequency ν, assuming that the displacement Δr_ν satisfies Newton's law:

$$m \frac{d^2 \Delta r_\nu}{dt^2} = e E_\nu e^{2\pi \nu i t} ,$$

with solution

$$\Delta r_\nu = -\frac{e^2}{m} \frac{E_\nu}{4\pi^2 \nu^2} e^{2\pi \nu i t} .$$

Since E_ν fluctuates, there will be a nonzero variance $\langle (\Delta r_\nu)^2 \rangle$. The S-state energy shift (9.1) is then obtained by summing over all frequencies. The result is the Lamb shift, due to vacuum fluctuations, and indeed shows that the S-level has a higher energy (by around 1000 MHz) compared to the $2P$-level.

9.3 Fluctuation Force

A good time before the idea that quantum fluctuations give rise to "forces" was being entertained and made explicit, there was interest in statistical forces, entropic forces, or in thermodynamic forces more generally. The terminology is not fixed, and various meanings are attached to these words.

In general, entropic forces can be viewed as forces driving an increase in entropy. The tendency of an undriven closed and isolated macroscopic system to maximize its entropy has measurable effects. It is a real thing. The expansion of a gas is not driven by energy—even an ideal gas may do it—it is driven by entropy. We say that the pressure of a gas has an entropic origin. Another example is the elasticity of a freely-jointed polymer. For entropic reasons, there is a force that tends to collapse a freely-jointed chain. This is often illustrated by the experiment where heat is produced in a rubber band when it is stretched. On the other hand, heating the rubber band contracts it. An entropic force is thus nothing more than a force caused by big numbers, felt mechanically, but of thermodynamic nature.

It is therefore reasonable to identify an entropic force with a statistical force, although the latter is more general. Suppose we have a probability density $\rho_X(y)$ for the positions y of particles in a macroscopic system (medium) when a slow probe is at position X. On a more microscopic level, there may be an effective interaction potential $U(X, y)$ between the probe and the medium particles, and thus a statistical force

$$F(X) = -\int dy \, \frac{\partial}{\partial X} U(X, y)\rho_X(y) ,$$

in one-dimensional notation. When

$$\rho_X(y) = \frac{1}{Z_X} \exp[-\beta U(X, y)]$$

is the Boltzmann–Gibbs weight for thermal equilibrium at inverse temperature $\beta = (k_B T)^{-1}$, we find that the statistical force

$$F_{eq}(X) = -\frac{d}{dX} \mathcal{F}(X) , \qquad \mathcal{F}(X) = -k_B T \log Z_X ,$$

can be derived from the (here, Helmholtz) free energy $\mathcal{F}$. It is in this sense that thermodynamic potentials are truly *potentials* of (statistical) forces. This idea can of course be extended to nonequilibrium media, where the statistical force is caused by gradients in temperature or chemical potential, and produces transport of energy or particles.

Such entropic, statistical or, more generally, fluctuation-induced forces also act between molecules. These are the London dispersion forces, named after Fritz London (1900–1954). The London dispersion force is the *only* molecular force for atoms and nonpolar molecules, and therefore of immense importance for understanding their macroscosopic behavior. The attraction at large distances is caused by fluctuations in the electron distribution around an atom or molecule, creating instantaneous electric fields. While deep down of a quantum mechanical nature, the force can be viewed as arising from the formation of instantaneous dipoles that attract each other.

The London dispersion force between individual atoms and molecules is rather weak and decays quickly, in fact as $1/R^6$, with the separation R. Yet it matters, as the attractive contribution to the Lennard-Jones potential, while the repulsion part at short distances includes the Pauli exclusion. For example, F_2 has a much lower melting point than Br_2, because bromine has a larger polarizability due to having more electrons than fluorine. As the electron cloud of the molecule gets bigger, the melting point increases, indicating that stronger forces do indeed need to be broken between the molecules.

The larger class of intermolecular forces (including London, Debye, and Keesom forces) are called van der Waals forces. They include the (permanent) dipole–dipole, ion–dipole forces, and others. They are important for explaining the microscopic origin of the van der Waals theory discussed in Sect. 10.3. A pleasant book explaining van der Waals forces has been written by Parsegian (2006).

9.4 Casimir Effect

Matter can modify the fluctuations of the electromagnetic field, resulting in an attractive interaction. The histories of interatomic potentials and Casimir forces are therefore intertwined. Again, entropic or thermodynamic forces at finite temperature are mathematically often similar to zero-temperature quantum-induced forces, while the physical origin of the fluctuations (thermal versus quantum) may be very different.

Work by Hendrik Casimir (1909–2000) and Dirk Polder (1919–2001) in 1948 showed that, for large separations, an atom would be attracted to a wall by a force Casimir and Polder (1948), and they corrected the interaction potential between two particles with static polarizabilities. They took the usual van der Waals forces as a starting point and corrected them for retardation effects. Indeed, when molecules or atoms are further away, there is a "finite" time for the fluctuation at one end to be felt by the other.

In the same year, Casimir showed that these expressions could also be obtained by studying the change in the electromagnetic zero-point energy. Also that Casimir effect (Casimir, 1948) between conducting plates is then due to quantum fluctuations of the electromagnetic field, a prototype of the fluctuation-induced interaction.

For some formalities, we consider a box with perfectly conducting walls separated by a distance d; the other dimensions L are assumed to be large. We want to estimate the energy in the box, and compare it with the case for $d = \infty$. That energy is found by summing all possible modes. The zero-point energy in the box is

$$\text{Energy}(d) = \frac{\hbar c L^2}{\pi} \sum_{n=0}^{\infty} \int_0^{\infty} dx \int_0^{\infty} dy \left(x^2 + y^2 + \frac{\pi^2 n^2}{d^2} \right)^{1/2} ,$$

which is divergent and as such devoid of physical meaning. Nevertheless, as Casimir was able to show, the difference for different values of d (or the derivative with respect to d) does have a well-defined meaning. He introduced a physical cutoff in frequency space, ignoring the very short waves for which the plate is no obstacle. Formally then, the energy required to bring the plates from infinity is $U(d) = \text{energy}(d) - \text{energy}(\infty)$, where

$$\text{energy}(\infty) = \frac{\hbar c L^2}{\pi^2} \frac{d}{\pi} \int_0^{\infty} dx \int_0^{\infty} dy \int_0^{\infty} dz \left(x^2 + y^2 + z^2 \right)^{1/2} .$$

It turns out that the difference is finite and can be calculated to give the Casimir force:

$$F(d) = -U'(d) = -\frac{\pi^2 \hbar c}{240 d^4} ,$$

a long-range attractive force. In Casimir's own words:

> There exists an attractive force between two metal plates which is independent of the material of the plates as long as the distance is so large that for wave lengths comparable with that distance the penetration depth is small compared with the distance. This force may be interpreted as a zero point pressure of electromagnetic waves.
> — H. Casimir in Casimir (1948).

Experimental agreement was obtained only some 50 years later in work by Lamoreaux Lamoreaux (1997).

9.5 Frenesy

The calculations of fluctuation-induced forces, be it the Lamb shift or the Casimir force, typically use semiclassical input, even though the context and the origin are thought to be fully quantum. That is not so surprising, perhaps. For one, the idea of a "force" is classical, and when the formalism is extended to include force operators as in the Hellmann–Feynman theorem, it remains restricted to time-independent situations. Secondly, starting from the full quantum mechanical picture of electrons interacting with the electromagnetic field, calculations become unfeasible. All this has to do with the detailed understanding of coarse-graining in quantum mechanical systems and the calculation of effective forces. Clearly, this in turn requires a good way to connect the quantum formalism with the real world, where stuff moves. By extrapolation, the issue is related to the famous measurement problem in quantum mechanics, and depending on the answer, certain concepts become available or not.

Then, after finishing the book I still felt that something was not terribly clear. In particular, there was no real way of representing or understanding what is meant by movement or process in quantum mechanics. One could only discuss observation, [...]. I was primarily disturbed by the inability to conceive of motion at all.
— David Bohm in The New Scientist Interview, 11 November 1982, speaking about his thoughts after talking with Einstein about his book Bohm (1951).

The dynamical activity of the vacuum is a quantum phenomenon that is discussed above for some pioneering experimental work in the context of quantum electro-dynamics. It also exists in quantum chromodynamics, however, although it proves more difficult to get good theoretical results there. There are other domains of appli-cation, even if sometimes only in spirit. As an example, it appears that the Nernst postulate, setting the scene and constraints for thermal properties near absolute zero, has a dynamical version that requires for its validity the presence of sufficient zero-temperature dynamical activity Khodabandehlou et al. (2023).

In even broader circles of nonequilibrium physics, it has been emphasized that terms involving the dynamical activity correct the standard fluctuation–dissipation relation Maes (2020). This is where the concept of *frenesy* is introduced to sum-marize the nondissipative kinetic aspects of a process Maes (2020), Maes (2018). It complements the standard entropic considerations, which, under Nernst –Planck conditions for the validity of the third law of thermodynamics, deals with an entropy that goes to zero anyway. Of course, one could also use the words *quantum noise* to emphasize that things keep changing, even in the ground state. For example, recall how, for the Dirac electron of Sect. 6.4, the chirality changes between the two mass-less Weyl electrons, following a Markov process. More generally, as we learn in Dürr et al. (2003), Tumulka (2022), interactions in quantum field theory can be described in terms of (stochastic) jumps corresponding to annihilation and creation events.

While the generalizing thoughts of the previous paragraph remain speculative, experimental work at very low temperatures continues to produce surprises. "Ther-malization" experiments near zero temperature would be nondissipative, and hence not seen as sliding down a (free) energy landscape, and solely driven by quantum noise.

Chapter 10
Phase Transitions

In his influential paper "More is Different" (Science, 1972) Anderson (1972), Philip Anderson (1923–2020) reminded us, students of Nature, that the global behavior of systems consisting of many particles may be very different, and indeed almost incomparably so and in unexpected ways, from the elementary motions of the individual particles. An anthropomorphic analogy can be found if we think of all the music festivals last summer: a crowd of people reacts differently than a single person. Or again, if we think about the pandemic and the way an infection can spread over the world, or not, depending on small changes in individual behavior. With birds or fish, we see flocking, and who could deduce the sociological organisation within an anthill by looking at a single ant? Evolution was not predicted by biochemistry, and it is hard to predict life from chemistry.

Keeping to physics, it is quite remarkable that large collections of molecules taken together show sharp transitions between very different macroscopic realizations. Compare liquid water with ice or steam, or a superconductor with a normal conductor, a ferromagnet with a paramagnet, and so on.

One rational approach to the general problem of collective phenomena is to consider the power of large numbers. In moving from microscopic laws to macroscopic behavior, cumulative and cooperative aspects exhibit the possible behaviors which are overwhelmingly typical given the imposed or intrinsic constraints. The word "typical" must be understood as the behavior that follows from the law of large numbers, introduced in (4.1). Big sums of local quantities, involving a large collection of particles in a macroscopic region, show little variance. This is a key result in probability theory. The rest is just fluctuations, large and small, with their own structure and laws— such as the central limit theorem in (4.2) and (4.5). These are also the main probabilistic ingredients for statistical mechanics, giving us the formalism to address complex, disordered systems or to move between various levels of description; see Bricmont (2022).

C. Maes, *Facts of Matter and Light*,
https://doi.org/10.1007/978-3-031-33334-7_10

Our motto has then become that the whole is more than the sum of its parts, not because the counting is wrong or more complex, but rather the opposite: a sum of many things often simplifies. After appropriate rescaling, only a few but very different results are possible, for example, finite versus infinite, positive versus negative, real versus imaginary, etc. And sometimes, most interestingly, a slight change in the constraints can give rise to drastic differences in the behavior of the sum, and it may correspond to a *critical* phenomenon, i.e., a very sensitive one indeed.

The term we use for the appearance of new or unexpected structures is "emergence." Something is emergent if it appears as a wholly new or unexpected behavior, apparently independent of the direct microscopic assumptions. The fact that it looks irreducible does not mean that it is underivable. Indeed, it becomes a fascinating program to derive it, much in the spirit of the natural philosophers since ancient times: How should we make the transfer from the microscopic laws governing microscopic constituents to macroscopic behavior? How should we connect a simpler, unifying microscopic description with the macroscopic diversity?

The science of emergence is statistical mechanics. It does indeed provide a way, a statistical way, of moving between different levels of description. Nature is hierarchical to an interesting extent, and depending on scales of length, time, or energy, new phenomena may appear. Statistical mechanics shows how to go from microscopic laws to mesoscopic and macroscopic behavior. It applies to nanophysics (and even much smaller scales) right through to cosmology.

There are various examples of emergence in physics. Below, we discuss the example of phase transitions. Other famous examples are the arrow of time (how to solve the irreversibility paradox[1]) and the emergence of chance (as a result of deterministic chaos). Emergence is also a theme in cosmology, considering for example the emergence of spacetime or of Lorentz covariance. Today, there are debates about the possibility of the (ontological versus epistemological) emergence of consciousness and free will. The latter problems appear to be of a totally different order of complexity. Maybe they are so magnificently different, at least from the simple examples in physics, that even more than the usual *more* is needed.

At phase transitions, new and unexpected macroscopic realizations appear for the same substance, depending on the imposed temperature, pressure, or density. You can walk on ice, but not on liquid water; yet, microscopically it is the same "H_2O". Likewise with magnets; they can become permanent, and then demagnetize at higher temperatures. Or when applying a voltage, the resistivity in a material may suddenly disappear as the temperature falls, while the microscopic aspects of the material remain unchanged.

The physics question is to understand how the same matter, with the same microscopic constituents and the same microscopic interactions, can support a diversity of exclusive physical properties and diverging functionalities when probed on the

---

[1] and why birds can fly... See d'Alembert's paradox (1752) for a rigorous proof that the drag force is zero for incompressible and inviscid potential flow, when a body moves with constant velocity relative to the fluid. Today, some call it an *information* paradox.

macroscopic scale. That has not always been so clear, and our understanding was still progressing very slowly in the first half of the twentieth century.

Another surprise was to follow, and another question was added to the one above, regarding the miraculous features of universality. At some points in the phase diagram, critical phenomena may arise in which a larger symmetry is so constraining that various quantitative aspects remain unaltered as our considerations move, say, from magnets to water. Not only can the same substance show very different behavior and properties, but some of that pattern is even—in some respects, quantitatively—independent of the substance. Such universality, occurring at critical points in equilibrium systems, allows classifications beyond the mere appearance of matter.

10.1 The Dream of Anaximenes

In Sect. 4.2, we mentioned the atomists, natural philosophers who wanted to explain Nature through Nature. One of these was Anaximenes. This pre-Socratic philosopher from more than 2500 years ago, already had such a project, or perhaps we should say a speculation or a belief. Anaximenes believed in the existence of unifying and overarching principles governing all life and matter, where appearance may change from microscopic to macroscopic.

Anaximenes' greatest influence is not through his theories of matter but originates instead in his ideas and ambition. It was not so very important that he thought of "air" as the building material for making everything; much more important was his idea that a substance is capable of changing form and characteristics. He firmly believed in phase transitions, and his "theory" was the first of its kind.

An important insight is that these phase transitions do not need to correspond to changes in the microscopic reality: steam and water do not look different on a microscopic level; they remain the same particles interacting with each other in a similar way. On microscopic scales, there is no real change, but the appearance will still be completely different depending on the pressure and temperature. Today, we understand melting and boiling, or *vice versa*, the way vapor condenses into rain or even hail and snow, even if we are always dealing with the same particles at a more microscopic level. In that sense, Anaximenes had the right intuition: different substances are condensations or dilutions of the same building blocks.

Nevertheless, the transition between different macroscopic phases, such as between ice, water, and vapor or between a normal and a superconducting material is proverbially miraculous.

In 1748, David Hume (1711–1776) discussed the water-and-ice example in his chapter on miracles in "An Inquiry concerning Human Understanding," (a version of this story had already been written by John Locke though): A Dutch ambassador tells the king of Siam (Thailand) about frozen water, water that is so strong that it can bear the weight of an elephant. The king is sure that this story is a lie; for him to call the water in a lake "strong, to support an elephant" was absurd, because it had never before been seen. Most likely, the Dutch ambassador lost all credibility.

Indeed, a full morning of a conference at the van der Waals memorial meeting in 1937 was spent discussing the possibility of a statistical mechanical explanation for the existence of phase transitions, and more precisely the question of whether one and the same system, microscopically defined in terms of its constituents and mutual interactions, could be in different macroscopic phases depending on its volume, energy, and particle number, or, in terms of intensive quantities, depending on its pressure, temperature, and density. More theoretically, the question was whether the partition function contained the right information to describe a sharp phase transition. As the debate was inconclusive, the chairman Hans Kramers (1894–1952), put the question to a vote and the *ayes* and the *nays* came out about evenly divided. It was still far from clear how to build a theory of phase transitions.

What makes these changes between the states (or phases) of matter even more incredible is that they are very sudden. They occur at specific values of pressure and temperature and bring about a drastic and immediate change in many visible aspects of matter, such as its density or internal order. An important clue derives from Gibbs' thermodynamics and its variational principles, We know that, depending on the precise balance between entropy and energy, the appearance of matter can differ greatly: a gas, for example, is much more disorderly than a liquid and will be the typical condition when the internal energy can be ignored (such as at high temperatures).

10.2   Percolation

The theory of phase transitions was one of the most important themes in the equilibrium statistical mechanics of the twentieth century. The central topic was the microscopic and statistical understanding of the location and nature of transitions between different macroscopic regimes. In the second half of the twentieth century, the theory of phase transitions reached out to different scientific fields and introduced concepts such as long-range order, symmetry-breaking, critical phenomena, and universality.

The physics of phase transitions took a huge step forward with the study of the so-called Ising model; see Wolf (2000) and the more recent review for mathematicians in Duminil-Copin (2023). This model provided a simplified picture of the transition between paramagnetism and ferromagnetism, and in fact the first proof of a phase transition. The key argument was provided by Rudolf Peierls (1907–1995) in 1936, and a detailed rigorous understanding was obtained for two dimensions in the work of Lars Onsager (1903–1976) in the years 1940–44. In what follows, we will briefly discuss some quantum examples which illustrate the emergence of macroscopic quantum coherence in sensational ways. We start, however, with what can be called a geometric analogue to (and is sometimes even said to be dual to) thermal phase transitions, viz., percolation.

The percolation transition is the mother of all phase transitions, in the same sense that space carries matter.[2] Percolation is the phenomenon in which a macroscopic path typically remains available to transport or flow. Think of an electricity distribution network used to conduct electricity from one end of the country to another; or a mask, a gas mask for example, or some other porous medium, designed to prevent certain particles from getting through; or an infectious disease that we want to contain using sanitary measures, to avoid the prospect of an epidemic.

On a more abstract level, we consider a graph (e.g., a regular square lattice) where nodes (vertices or sites) are connected to other nodes, making bonds (edges). Each bond is *open* with probability p and *closed* with probability $1 - p$, independently of all the others. This value of $p \in [0, 1]$ is thus a parameter giving the density of open bonds; see Fig. 10.1.

Let us make a model and apply some simple reasoning (see, e.g., Grimmett, 1999 for more details). Think about a large number of bonds, $2N(N - 1)$ for a square lattice of size $N \times N$ nodes, each having 4 nearest neighbors (except at the boundary). In other words, each node has 4 different bonds attached to it. Now we start somewhere in the center of the square and try to walk via open bonds to the boundary. If the boundary is far away, for large N, it is not obvious that we will succeed (see Fig. 10.1). What exactly is at stake here? Any fixed path (that does not cross itself) involving n bonds has a small probability of being totally open, in fact scaling as p^n, which is exponentially small for large n. Therefore, percolation along any given path is unlikely. That is why in one dimension, for a chain of N nodes, each with two neighbors, there is no percolation except, trivially, when $p = 1$. But in two (and higher dimensions) something can save us: there are many possible paths. The number of possible paths does not depend on p of course; it is purely geometric. In fact, in two dimensions, the number of self-avoiding paths of length n scales as 3^n (more or less). That is, when we arrive at a node, we can go to three other nodes (excluding the one we just came from). This repeats itself at each step, giving $3 \times 3 \times \ldots \times 3 = 3^n$ possible ways. And that 3^n must be compared with the *cost* of p^n. Clearly now, at least informally, when p is (too) small, $p^n \times 3^n$ will also be too small to allow percolation. When p is large enough, however, and still $p < 1$, there is a possibility of percolation (see Fig. 10.2). Indeed, that is what happens: for bond percolation on the square lattice, the threshold is $p_c = 1/2$. We think here in the limit $N \uparrow \infty$: when $p \leq 1/2$, there is typically no percolating path (infinite open cluster), while for $p > 1/2$, there is.

For large dimension d of the regular lattice, as follows from the arguments above, $p_c \approx 1/2d$.

We have a (percolation) transition at a threshold p_c. The point is that the transition is there and it is sharp. For $p \leq p_c$, there is no open macroscopic cluster, and for $p > p_c$ there is. The sharpness of that transition also follows heuristically from the

---

[2] The analogy is perhaps deeper than at first sight. We know from the work of Kees Fortuin and Piet Kasteleyn (1924–1996) how Potts models, for example, may be represented and have a dual description as a "random cluster" percolation model Georgii et al. (2023). The latter is a fluctuating geometry in the connectivities of which the correlation functions of the Potts model can be expressed at all coupling strengths.

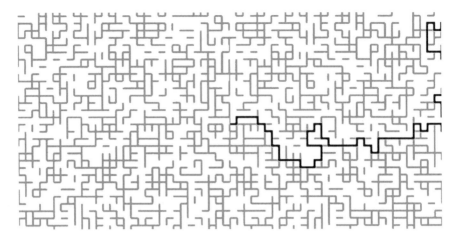

Fig. 10.1 A random configuration of open and closed bonds, with an open path reaching the boundary

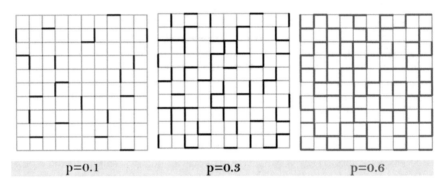

| p=0.1 | p=0.3 | p=0.6 |

Fig. 10.2 Representations of typical bond configurations for a square lattice for three values of the density p. The first two are subcritical, and the one on the right is supercritical

arguments above: *two things matter*, the cost p^n and the number c^n of paths of length n (where $c \approx 2d - 1$). The sharp transition is at the point where there is a precise balance. If not exactly in balance, either non-percolation or percolation wins, thus marking a sharp transition between two macroscopically different phases. The mathematical arguments are needed as the above is not always true for other lattices, and the proofs for when it does hold, are only recent.

The ideas above are of much wider validity and can be made rigorous. In thermodynamic jargon, the macroscopic phase that wins is the one with the lowest free energy. The latter is an expression of the competition between energy (cost) and entropy (degeneracy). Points of phase transition are determined by the possibility of having various phases that minimize the free energy. Those are such powerful ideas that they survive the transition from geometric variables to spin variables, from spins to fields, and from classical systems to quantum systems.

10.3 Criticality and Universality

Can we liquefy a gas by increasing the pressure? The answer is, not always. It must first be cooled down below its critical temperature. The foundational experiment on the condensation behavior of gas was devised by Thomas Andrews (1813–1885) in 1863.

He used carbon dioxide as the test gas. Above 50 °C it behaves as an ideal gas. This means that the isothermal curves obey the ideal gas law discovered by Boyle (1627–1691), Gay-Lussac (1778–1850), and Mariotte (1620–1684). Ideal gases do not condense. Their constitutive particles can be treated as point-like, having no volume and no interactions as far as their equilibrium properties are concerned. We expect condensation to require interactions, e.g., we assume that the atoms or molecules repel each other when they get truly close, as happens with colliding hard spheres, and that they attract each other when far way. In Andrews' experiment (see Fig. 10.3), a drastic change happens at the so-called critical temperature for carbon dioxide, which is 30.9 °C. For lower temperatures (only), carbon dioxide gets transformed from any of the states we usually call liquid to any of those we usually call gas, without losing homogeneity Andrews (1869). We have to "turn around" the critical point, as indicated for the phase diagram of water in Fig. 10.4.

We do this by heating, then changing the pressure, and then cooling down again. The concepts of critical temperature and critical pressure were thus established. Carbon dioxide has a critical pressure of 73 atm. At the critical point, gas and liquid coincide, which is different from being in equilibrium, as at a phase transition.

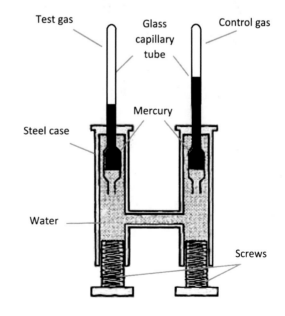

Fig. 10.3 Andrews' experiment. A thick-walled glass capillary tube containing a gas is enclosed with a drop of mercury. The capillary tube is sealed into a metal compression chamber. The chamber contains water and is connected to an exactly similar chamber with a glass tube of the same type but containing air. The latter tube serves as a manometer. The experimental gas is contained in the capillary tube. The pressure applied on the water by screwing in the screw is hydraulically transmitted by the water to the volume of air. The upper part of the carbon dioxide tube is surrounded by a heat bath so that the temperature can be varied

Fig. 10.4 Phase diagram of water. The lines separate phases (solid ice, liquid water, and vapour). The triple point is the location where all phases coexist. The critical point marks the end of the liquid–gas separation

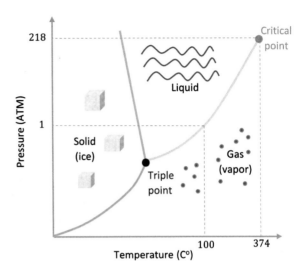

That means that we should improve on the ideal gas law by taking into account changes in pressure and volume. For those theoretical developments, one of the most deserving pioneers is Johannes Diderik van der Waals (1837–1923), recipient of the 1910 Nobel Prize in Physics, who in his PhD *About the Continuity of the Gaseous and Liquid States* of 1873 had already derived an equation of state in which both the gas and the liquid state ensue from the same microscopic considerations. Maxwell was so impressed that he thought it worthwhile learning Dutch in order to read the work of van der Waals in its original language[3]:

> [...] that there can be no doubt that the name of van der Waals will soon be among the foremost in molecular science. [...] It has certainly directed the attention of more than one inquirer to the study of the Low-Dutch language in which it is written.
> —James Clerk Maxwell (Nature).

Van der Waals suggested the equation

$$\left(P + a\frac{N^2}{V^2}\right)(V - Nb) = Nk_{\mathrm{B}}T .$$

There are two parameters, a and b. The latter takes into account the particle volume; there are N particles of volume approximately b. The pressure becomes infinite when the volume V decreases to Nb. Secondly, the parameter a takes into account the intermolecular interactions. Remember that the pressure is the derivative of the energy with respect to the volume:

$$a\frac{N^2}{V^2} = -\frac{\partial}{\partial V}E_{\mathrm{intermolec}}, \quad E_{\mathrm{intermolec}} = aN\frac{N}{V} .$$

---

[3] A translation became available only in 1988.

Fig. 10.5 Van der Waals isotherms. At high temperatures, the Ideal Gas Law holds true. Below a critical temperature, the possibility of a phase transition arises. There is a jump in density and entropy at the liquid–gas transition

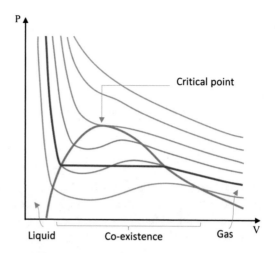

where we indicated how the energy may be thought to depend on the density. For small molecules like nitrogen or water, we have $b^{1/3}$ of the order of a couple of angstrom units, while b/a can be a few eV. For helium, a noble gas and close to ideal, a is much smaller ($a \simeq 3.46 \times 10^{-3}$ Pa m^6).

In Fig. 10.5, we see the pressure as a function of the volume, giving the isotherms

$$P = \frac{Nk_B T}{V - Nb} - a\frac{N^2}{V^2} . \qquad (10.1)$$

The first term on the right-hand side is the Ideal Gas Law for the effective volume. The second term decreases the pressure by the intermolecular forces.

At high temperatures or for large volumes we get isotherms like those of an ideal gas. At lower values of T, the behavior gets more interesting. The fact that the pressure can increase at lower volumes is something of an artifact here and can be remedied by the so-called Maxwell construction (which we skip). The main point is that, when the pressure decreases, there is an abrupt increase in volume: this is a phase transition from the liquid to the gas phase. In the same Fig. 10.5, for the same fluid, we see a transition between the liquid and the vapour phase: a miracle explained.

And that is not all. In 1880, van der Waals also formulated the law of the corresponding states in which a first sign of universality arises: the equation of state can be written in a universal form, independently of the specific substance.

The critical point is the unique point on the van der Waals isotherms where both the first and second derivatives of the pressure with respect to V are equal to zero. A little calculation starting from (10.1) gives, at the critical point,

$$V_c = 3Nb , \quad P_c = \frac{a}{27b^2} , \quad k_B T_c = \frac{8a}{27b} .$$

Be ready for the second miracle. Rewriting the van der Waals equation in terms of the reduced variables

$$t := \frac{T}{T_c} , \quad p := \frac{P}{P_c} , \quad v := \frac{V}{V_c} ,$$

we get an equation of state that no longer depends on the (fluid parameters) a or b:

$$\left(p + \frac{3}{v^2} \right) \left(v - \frac{1}{3} \right) = \frac{8}{3} t .$$

For given t and p, v is determined independently of the substance. This was the first example of universality; after rescaling, the gas-liquid transition at the critical point is the "same" for all fluids.

At that point, universal properties appear, and details can be scaled away. Scaling behavior refers to the disappearance of intrinsic space or time scales. Whether you look at the system from close by or from afar, with zoom-in or zoom-out, the mutual correlations between the components of the substance are invariant.

10.4 Superconductivity

There are materials that exhibit zero electrical resistance and expel magnetic fields below a characteristic temperature. They are then superconducting.[4]

The history of superconductivity began with the search for a good thermometer and more broadly the exploration of physical phenomena at low temperatures. Low-temperature physics became possible thanks to ever-improving ways to cool matter. Temperatures got lower and lower, but good thermometers were also needed. One idea to measure the temperature was to measure the resistance through a wire, for the resistance depends on the temperature. A (low) temperature could be inferred by measuring the resistance. And that is how in 1911 Heike Kamerlingh Onnes (1853–1926) discovered superconductivity in mercury.[5]

A necessary ingredient in the discovery of superconductivity was the liquefaction of helium. The temperature required to produce liquid helium is low because of the weakness of the attractions between the helium atoms. This was also achieved by

---

[4] There exist in fact different types of superconductors, depending on the critical field(s) above which superconductivity is lost and/or the magnetic field is expelled.

[5] Kamerlingh Onnes discovered superconductivity with Gilles Holst (1886–1968), his PhD student, inventor of the low-pressure sodium lamp who would later become director of Philips. Holst did the experiment and is thanked in the acknowledgments. For more background about the Dutch school, see Sengers (2002), in which Anneke Sengers narrates the story of understanding phase separation and fluid mixture criticality, and details the collaboration between Johannes Diderik van der Waals and Heike Kamerlingh Onnes.

Kamerlingh Onnes, in 1908, at Leiden University. He produced liquid helium, going below the boiling point of 4.2 kelvin at atmospheric pressure.

Kamerlingh Onnes and Jacob Clay (1882–1955) then set out to redo earlier experiments by Dewar on the reduction of resistance at low temperatures. They started with platinum and gold, but Onnes replaced these later with solid mercury, also a metal and more readily refinable. Of course, Onnes used liquid helium as a refrigerant. On April 8, 1911, Onnes noted that the resistance of mercury went to about zero: at 4.19 K, the resistivity suddenly disappeared, Kamerlingh Onnes (1911a).

In the last 1911 paper "On the sudden change …," Onnes explains how the "specific resistance" became thousands of times smaller relative to the best conductor at ordinary temperature. When Onnes later reversed the process, he found that, at 4.2 K, the resistance returned to the material.

For about 40 years, Kamerlingh Onnes served as professor of experimental physics at the University of Leiden, having a great and immediate influence on low-temperature physics in the lowlands.

About 75 years later, J. Georg Bednorz and K. Alex Müller discovered superconductivity in a lanthanum-based cuprate perovskite material, Bednorz and Mueller (1986). This has a transition temperature of 35 K, and so became the first of a whole new class of what would be called high-temperature superconductors. Ching-Wu Chu replaced the lanthanum with yttrium, and the critical temperature rose to 92 K, Wu et al. (1987). Now liquid nitrogen could be used as a refrigerant at atmospheric pressure, which is very convenient for applications.

Many other cuprate superconductors have since been discovered, but room-temperature superconductivity is not here yet. The theory of superconductivity remains one of the major outstanding challenges of condensed-matter physics.

10.5 Superfluidity

If a liquid is able to flow through narrow channels without apparent friction, we call it a superfluid. It flows without resistance. There are usually other interesting properties that come with this. For example, a superfluid liquid in a bucket will stay at rest even when we rotate the bucket (Hess–Fairbank effect). The superfluid can penetrate fine pores that are closed to ordinary liquids and to many gases. It can also creep upward along walls and make an excellent heat conductor. A superfluid liquid can flow forever down a narrow capillary without apparent friction, in the same way as a current in a superconducting ring can flow forever. This brings us to the physics of quantum liquids.

As we saw a moment ago, helium-4 was liquefied in 1908. Kamerlingh Onnes used liquid hydrogen as an intermediate step. An old friend of Rutherford, John McLennan (1867–1935) built the second laboratory in the world to successfully produce liquid helium in 1923. Much as a result, between 1923 and 1933, the Leiden and Toronto laboratories dominated research in low-temperature physics using liquid helium. In April 1934, Pyotr Kapitza (1894–1984) working at the Cavendish in

Cambridge under the supervision of Rutherford since 1921 and developing there a new kind of liquefier based on gas expansion, produced liquid helium. From then, many other laboratories were able to produce liquid helium as well, giving a boost to low-temperature physics. While Kapitza was the driving force in the low temperature lab in Cambridge, in September 1934 he was "detained" in Moscow during one of his regular visits to his family in the Soviet Union. John Allen (1908–2001) and Don Misener (1911–1996) were attracted from Toronto to continue the low-temperature physics research in Cambridge.

In 1936 and 1937 it was found that, below 2.17 K, the thermal conductivity becomes very high. This points to the possibility of magnificent convection with a very low viscosity. This low-temperature phase is now known as He-II, and the transition happens at the so-called lambda point. All this was discovered around the same time in 1938 by Pyotr Kapitza in Moscow and by John Allen and Don Misener in Cambridge UK, Kapitza (1938), Allen and Misener (1938). They measured the viscosity of the helium contained in a thin tube as a function of temperature, and both groups found a discontinuous drop at the lambda point.[6] Kapitza coined the term superfluidity for this behavior.

Today, superfluidity is directly observable in helium isotopes and in ultracold atomic gases. Some believe it also prevails in neutron stars.

The superfluid phase of helium-4 is an excellent coolant because of its extremely high thermal conductivity. It is used for high-field magnets, with applications in experimental high-energy physics. Helium has two stable isotopes. Helium-4 has an integer (whole-number) spin while helium-3 has a half-integer spin. Helium-3 is a liquid right down to absolute zero. It is rarer but, more importantly, the helium-3 atom is a fermion. It was therefore interesting news that, at very low temperatures (a few thousandths of a degree above absolute zero), helium-3 also becomes superfluid. The reason is that the He-3 atoms form two-atom Cooper pairs that are bosonic and condense into a superfluid. In 1972, Douglas Osheroff, David Lee, and Robert Richardson verified the superfluid properties of He-3, Osheroff et al. (1972).

10.6   Bose–Einstein Condensation

Electric lamps emit light in all directions and with all colors. Around 1917, Einstein showed how atoms under certain conditions can be stimulated to emit light. Moreover, this emitted light has a wavelength (color) and direction that unambiguously matches the light that causes the emission. The amplification of light by light became known by the name of LASER, short for *Light Amplification by Stimulated Emission of Radiation*. We can also ask whether lasers can be made from matter. In other words, is it possible to bring atoms together in a state where they form one coherent assembly.

---

[6] John Allen used a movie camera to film his experiments, such as the superfluid helium fountain. This was one of the first videos of experimental physics, a great legacy available to the general public.

In the early 1920s, Satyendra Nath Bose (1894–1974) assumed certain rules for photon counting. These rules give rise to a statistics, now called Bose statistics, which was seen by Einstein to give the possibility of condensation. In the paper Einstein (1925) written in 1924, Einstein generalized the theory to certain types of particles with nonzero mass, so-called massive bosons. The prediction was that, when a gas is made of such particles and cooled to very low temperatures, the particles would collect in the lowest possible energy state. This is referred to as Bose–Einstein condensation. In this condition, all atoms are described by the same wave function; they overlap to such an extent that they form one superatom. "But, this appears to be as good as impossible," reads the last line of Einstein's paper.

In 1995, this phenomenon was observed experimentally, Anderson et al. (1995). A Bose–Einstein condensate was produced by Eric Allin Cornell and Carl Wieman, containing 2000–3000 rubidium atoms at a temperature of 20 nanokelvin (nK). Note that only a few months earlier, the same group had broken the temperature record in atom–cooling when reaching 200 nK, made possible by a new type of very confining potential trap to allow efficient evaporative cooling. Cornell had just obtained his doctorate and joined Carl Wieman at the University of Colorado Boulder as a postdoc. The project was a laser cooling experiment. During his two years as a postdoc, he came up with a plan to combine laser cooling and evaporative cooling in a magnetic trap cold enough to create a Bose–Einstein condensate. Rubidium atoms have light-absorbing and magnetic properties that make them easy to manipulate. Laser beams bounced photons off the atoms to slow them down. Next, as each atom is a tiny magnet, external magnetic fields were used to cool the atoms further and to keep them packed together tightly enough for their wave functions to overlap. A radio-frequency oscillating magnetic field was tuned to resonate with the higher energy atoms, which caused hot atoms to flip, reversing their magnetic orientation so that they were ejected from the trap. And they got what they wanted: a dense, light-reflecting core of 3000 overlapping rubidium atoms at the center of a small bottle.

At the same time Wolfgang Ketterle was doing similar work, but with sodium atoms, David et al. (1995). In each case, the condensate was shown to consist of fully coordinated atoms, very similar to the light packets in a laser.

Deborah Shiu-lan Jin (1968–2016) and her team produced the first fermionic condensate in 2003. This was a new form of matter. Fermionic atomic gases had to be cooled to less than 100 billionths of a degree above zero. This led to the formation of a molecular Bose–Einstein condensate.

Chapter 11
Nonlocality: Spooky Action at a Distance

Let me start with what is obvious. Quantum mechanics was part of the revolution in twentieth-century physics. It has pervaded the whole of physics, with quantum effects in all branches of physics, becoming testable and visible whenever the relevant range of lengths, times, and energies is reached. The quantum revolution has affected most of chemistry too, for obvious reasons, e.g., providing our understanding of the chemical bond and allowing us to calculate electronic contributions to chemical properties. And there are many applications to medicine and engineering. The entire computer industry is built on quantum mechanics. All our devices (desktop computers, laptops, tablets, smartphones) and many household appliances are governed by computer chips, themselves made possible thanks to our understanding of quantum physics. When we use a laser, to make a phone call or scan a label, we use quantum physics. Quantum technology in the form of lasers, transistors, and clocks has reached every aspect of our daily lives. In turn, they are invaluable for progress in experimental physics. In a word, for our physical understanding and ability to manipulate Nature, quantum mechanics is now of central importance.

Less obvious, surprisingly, is to agree on what quantum theory ought to be about. Various "versions" of quantum mechanics have appeared, and the jury is still out to decide what empirical tests might distinguish them. We should probably start by agreeing on what we agree about, and that may well be a summary of systematic departures from classical mechanics that are manifested through quantum mechanics:

- Its radical particle point of view, where quantization is discretization, elucidating the corpuscular nature of light and interaction, etc. This is responsible for the fundamental changes in low-temperature calorimetry, the discrete nature of spectral lines, the photoelectric effect, the Compton effect, and so on. In this way, quantum mechanics realized the old idea of atomism and opened the way to elementary particle physics. However, the mechanics of these "atoms" departs drastically from that of Newton. Such ideas were discussed in Chap. 5.
- The very nonclassical one-particle behavior allowing tunneling, i.e., getting past potential barriers, important for decay, scattering, and radiation processes. The

motion of particles shows wave-like properties, e.g., via interference phenomena (see also Chap. 6).

- The presence of discrete inner degrees of freedom that interact with electromagnetic fields, via the so-called spin, whose origin is, however, (partially) relativistic. We discussed this in Sect. 5.6.

- New properties of matter, such as magnetism, superconductivity, superfluidity, and new collective behavior in general. The deviation from Maxwell–Boltzmann statistics, realizing Bose–Einstein and Fermi–Dirac distributions. Think of ultracold gases with the phenomenon of Bose–Einstein condensation, or the theory of electron conduction, or the state of matter in neutron stars, etc. Some of this was dealt with in Chap. 10.

- Vacuum polarization and (ground state or vacuum) quantum fluctuations, showing dynamical activity in vacuum and in (zero) temperature ground states (see Chap. 9). Associated with these intrinsic quantum fluctuations there appears a fundamental quantum uncertainty or chance element.

- Its weird nonlocal multiple-body influences, where entanglement and many-body coherence reigns.

This chapter will focus on the last point, sketching the main ingredients of the weird behavior known as nonlocality, i.e., violation of the principle of "local causality." Strangely enough, the crucial quantum feature of entanglement, already discussed by Heisenberg, Schrödinger, and Einstein, played little or no role in much of the quantum developments of condensed matter or elementary particle physics before 1990. Many-body physics and the physics of strongly correlated particles really take off in fascinating directions, however, when entanglement gets involved. And more can be expected; see also in the next Chapter.

Insights into nonlocality first grew from thought experiments, enquiring into the meaning of quantum mechanics. From the very beginning, quantum mechanics has struggled with questions concerning its completeness and the notorious measurement problem. Thought experiments evolved from such considerations by the founders of quantum mechanics, including Schrödinger and Einstein. Quite remarkably and most interestingly, those musings have given rise to, and are still giving rise to real experiments. Moreover, it is certainly a pleasant side-effect of the 2022 Nobel Prize in Physics that this has revealed a clear causal connection between quantum foundations and quantum technology.

Two of the prizewinners, John Clauser, and Alain Aspect, did the first experiments with entangled photons, confirming the violation of the Bell inequality. These were established by John Stewart Bell (1928–1990) as a natural consequence of the Einstein–Podolsky–Rosen(–Bohm) thought experiment (of Sect. 11.3), which questioned the completeness of quantum mechanics as a description of Nature. Bell's theorem, from 1964, showing that there is no local description of quantum phenomena, was turned into a Bell test by Clauser, Horne, Shimony, and Holt in 1969, when they came up with a variant of Bell's theorem. The first Bell test then came in 1972, set up by Freedman and Clauser. Then, in 1981–1982, came Aspect with Grangier, Roger, and Dalibard who improved on those first Bell tests. The third 2022 Nobel prizewinner was Anton Zeilinger, who transferred the entanglement ideas to various applications in quantum information, communication, and cryptography, with quantum teleportation as one of the highlights.

11.1 About Alice, Living Far Away from Bob

An entangled system refers to a composite system, typically made up of individual particles that have previously been in interaction. Imagine now that after that initial preparation, they are being spatially separated, with two experimenters Alice and Bob handling them at those distant locations. However, before touching the particles in any way, they remain entangled, no matter how far apart they are. This enables nonlocal influences, as we will show.

Let us first be a little more formal about entanglement. The state as described by a wave function is entangled if it cannot be factorized into a product of individual particle states: so, for two particles with coordinates x, y, the wave function $\psi(x, y) \neq \phi_1(x)\phi_2(y)$. In vector (quantum state) notation, this is interpreted as follows. Consider the state

$$|\psi\rangle = \sum_{k,\ell} c_{k,\ell} |k, \ell\rangle, \qquad |k, \ell\rangle = |k\rangle \otimes |\ell\rangle \,,$$

in a two-particle composite system. The state is separable if $c_{k,\ell} = a_k b_\ell$; inseparable states are called entangled states. As a consequence, in those entangled states, one particle (or spin, etc.) cannot be meaningfully specified in its state without considering the other(s). Alice and Bob count on "meeting in the sum."

You know, I try, when darkness falls,
to estimate to some degree —
by marking off the grief in miles —
the distance now from you to me.

And all the figures change to words:
confusion, which begins at A,
and hope, which starts at B, move towards
a terminus (you) far away.

Two travelers, each one with a light,
move in the darkness, silent, dumb.
The distance multiplies all night.
They count on meeting in the sum.

— For schoolchildren, Joseph Brodsky (1940–1996). Translated by Tatyana Amelina and Harry Thomas.

Let us take two qubits, i.e., two two-state quantum mechanical systems (say, two spin 1/2 systems) referred to as subsystems A and B, with basis vectors $|0\rangle_A, |1\rangle_A$ and $|0\rangle_B, |1\rangle_B$, respectively. They are shorthand for the vectors $|1\rangle = (1 \ 0)^T$ and $|0\rangle = (0 \ 1)^T$ in $\mathbb{C}^2$, for each of the two components of $\mathbb{C}^2 \otimes \mathbb{C}^2$. An example of an entangled state (in the composite system) is

$$v = \frac{1}{\sqrt{2}} \left(|0\rangle_A \otimes |1\rangle_B - |1\rangle_A \otimes |0\rangle_B \right) .$$

Now, there is no definite pure state either for system A or for system B; the restriction to A is described by a density matrix $\rho = \frac{1}{2} \left(|0\rangle_A \langle 0| + |1\rangle_A \langle 1| \right)$. For example, the von Neumann entropy $S_{vN}(\rho) = -\text{Tr}[\rho \log \rho] > 0$ of the subsystems is greater than zero, a sign that they are entangled.

Let us now consider an experiment. Suppose Alice observes system A and Bob observes system B. Let Alice make a measurement in the eigenbasis $|0\rangle, |1\rangle$ of A. For example, she is measuring whether the "spin is up." This has probability $\text{Tr}[\rho P_1]$, for the projection $P_1 = |1\rangle_A \langle 1|$. She will measure a frequency of 1/2 for "up". In fact, the two possible outcomes "up" and "down" occur with equal probability. For the state v above, if A measures 0, Bob will certainly measure 1 on the other side, and if A measures 1, Bob will certainly measure 0 on his side. Formally, this can be seen from the conditional probability

$$\frac{(v, P_0 \otimes P_1 v)}{(v, P_0 \otimes \mathbb{I} v)} = 1 ,$$

where $\mathbb{I}$ stands for the unit matrix (on the second component).

However, if Alice measures the spin in another basis (even without telling anybody), then the resulting statistics for Bob will change: half the time he will measure 0 and half the time he will measure 1. For example, if P_s denotes projection onto the vector $\frac{1}{\sqrt{2}} (1 \ 1)^T$, given by $P_s = \frac{1}{2} (|0\rangle + |1\rangle)_A (\langle 0| + \langle 1|)$, then

$$\frac{(v, P_s \otimes P_1 v)}{(v, P_s \otimes \mathbb{I} v)} = 1/2 .$$

Therefore, what Bob measures is influenced "in flight" by the type of measurement Alice is making, and that remains true even if the systems A and B are spatially separated. This is quite some action at a distance. We will revisit this scenario in various forms in the following sections to understand its meaning.

Remember though that the outcome of Alice's measurement remains random (and specified by the density matrix ρ), so no information can be transmitted to Bob by acting on her system.

11.2 Einstein's Boxes

In the 1920s and 30s, Einstein was almost alone in realizing that quantum mechanics as it was understood at that time (and as it is explained even in recent textbooks) faced a problematic dilemma between completeness and locality. He simultaneously developed various thought experiments to make that point. One of the simplest (and clearest) is known as the Einstein's boxes thought experiment. See Norsen (2005), Bricmont (2016) for a useful summary and description. For aspects of its experimental realization, see Gea-Banacloche (2002).

A single particle is confined to a box B. After a while, a barrier is inserted to split the box into two half-boxes B1 and B2. They are then separated, at which point the boxes may be opened and their contents examined (see Fig. 11.1).

Einstein examined two possible ways of analyzing this experiment:

> The probability is 1/2 that the ball is in the first box. Is this a complete description?
> NO: A complete description is: the ball is (or is not) in the first box. That is how the characterization of the state of affairs must appear in a complete description.
> YES: Before I open them, the ball is by no means in one of the two boxes. Being in a definite box only comes about when I lift the covers. This is what brings about the statistical character of the world of experience, or its empirical lawfulness. Before lifting the covers the state [of the distant box] is completely characterized by the number 1/2, whose significance as statistical findings, to be sure, is only attested to when carrying out observations. Statistics only arise because observation involves insufficiently known factors foreign to the system being described.
> — Einstein in a letter to E. Schrödinger, 1935.

The YES view is what many followers of the Copenhagen interpretation of quantum mechanics would answer. But that, as Einstein saw, is incompatible with an assumption of locality or separability that "the second box, along with everything having to do with its contents is independent of what happens with regard to the first box (separated partial systems)." The NO view says that the wave function and its evolution from the linear Schrödinger equation is not complete. Physics should do better.

Let us add a formula. The quantum state of the particle is

$$\psi = \frac{1}{\sqrt{2}}(\psi_1 + \psi_2) \, ,$$

where ψ_i is completely localized in box Bi. Let us apply the collapse postulate (as the Dirac–von Neumann formalism prescribes). The wave function in B2 immediately drops to zero if and when we open the box B1 and we find the particle there. But if there is locality, the actual physics in B2 cannot have changed. Hence, the details of ψ_2 before the measurement do not seem to matter that much, which suggests that the quantum state description is incomplete. Under the assumption of locality, the description in terms of the wave function cannot be complete. See Fig. 11.1 for a similar thought experiment.

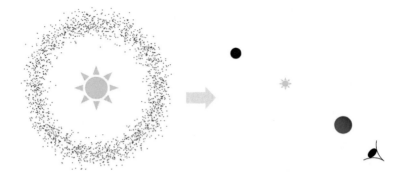

Fig. 11.1 Einstein's thought experiment, similar in style and message to the boxes and the EPR-B pairs. *Left* The total wave function of a particle pair spreads from the collision point. *Right* The observation of one particle collapses the total wave function. Was there a matter of fact concerning the second particle before measuring? If not, then there is a nonlocal influence. If yes, then the usual (textbook-) quantum description in terms of the wave function only is incomplete (but may still be nonlocal upon completion)

For completeness, we may mention that some (especially theoretical physicists when forced to speak out on interpreting quantum mechanics) claim to like "many-worlds" as a potential escape.[1] That perhaps is an escape into nihilism, of similar excitement as is the recourse to solipsism.

11.3   Einstein–Podolsky–Rosen Argument

The criticism that Einstein made regarding the state of quantum mechanics at the time had nothing to do with indeterminism or with the so-called statistical nature of quantum predictions. His main point was that he understood that orthodox quantum mechanics was incomplete (if local). The Einstein–Podolsky–Rosen (EPR) paper of 1935, *Does the Wave Function Provide a Complete Description of Physical Reality?*, sought to make that point Einstein et al. (1935). Schrödinger gave an extension of the EPR argument using maximally entangled quantum states, Schrödinger (1936). In the original 1935 EPR paper, the argument was formulated in terms of position and momentum (which are observables having continuous spectra). The argument was later reformulated by David Bohm in terms of spin. This "EPR-B" version is conceptually simpler and also more closely related to the recent experiments designed to test Bell's inequality. A reference article appeared in a selection of seminal papers

---

[1] It is often chosen for its minimalism, i.e., in having only the Schrödinger equation to deal with, while all the same "a zillion of worlds" are being added. It is also said to be the unique formalism dealing with (quantum) cosmology and to solve the measurement problem through the theory of consistent/decoherent histories, *quod non.*

of the first hundred years of *The Physical Review*, Goldstein and Lebowitz (1995). See also Bricmont et al. (2022) for a conceptually simple proof of nonlocality based on EPR-Schrödinger.

The main conclusion of the EPR paper is that, if we accept the basic principles of locality as we were very much used to throughout classical physics and perhaps throughout all natural sciences, then there must exist some kind of hidden variables; the description in terms of the wave function cannot be complete.

Remember that, in orthodox quantum mechanics (say as presented by Bohr, Heisenberg, and von Neumann), measurements do not reveal preexisting values. There is no question there about the value of no matter what observable before measuring it, and hence even a Copenhagen Laplacian demon would not and could not predict in advance which outcome would be realized for sure.[2] Nevertheless, the 1935 EPR paper points out that this orthodox point of view is in conflict with locality: if nonlocal influences are forbidden, and if certain quantum theoretical predictions are correct, then measurements whose outcomes are strictly correlated must reveal preexisting values. That is the reason why Einstein described orthodox quantum mechanics as "incomplete."

Despite the many excellent sources for understanding EPR, e.g. in Einstein et al. (1935), Goldstein and Lebowitz (1995), Bricmont et al. (2022), we still give a brief sketch of the main argument.

Here we can start from entangled pairs of photons or electrons. A pair of spin-1/2 particles is in the entangled spin singlet state

$$\frac{1}{\sqrt{2}}\left(|10\rangle - |01\rangle\right),$$

with 1 and 0 referring to spin up and down. The particles are sent in opposite directions. The measurement of no matter what spin or chirality by Alice to the left instantaneously "influences" the measurement result Bob will obtain to the right, in such as way as to get perfect (anti)correlations. That is the prediction derived in quantum mechanics.

Each spin measurement on one of the particles gives a result of either "up" or "down." Measuring the spin of both particles along some given axis shows perfect anti-correlation as predicted by the quantum formalism. Now think about simultaneous measurements on two such spatially-separated particles. Locality[3] requires that any disturbance triggered by the measurement on one side cannot influence the result of the measurement on the other side. But the only way then to ensure the perfect anti-correlation between the results on the two sides is to have each particle carry a preexisting determinate value (appropriately anti-correlated with the value carried by the other particle). We might think of appealing to indeterminism for a way out,

---

[2] Bohr would say that the observer co-creates physical reality by the questions he poses in his experiments.

[3] Einstein's—or everybody's—*Trennungsprinzip*, as he put it in his 1935 correspondence with Schrödinger.

but in fact any element of locally-confined indeterminism would at least sometimes spoil the predicted perfect anti-correlation between the outcomes. Preexisting values are thus the only local way to account for perfect anti-correlations in the outcomes of spin measurements along identical axes.

We will see below in the discussion on the Bell inequality how the existence of such preexisting values (as a consequence of the assumed locality) is in conflict with other observations and experiments. The final conclusion is then mind-boggling: all that remains is to give up (some form of) universal locality.

> A doubt which makes an impression on our mind cannot be removed by calling it metaphysical.
> — H. Hertz, Die Prinzipien der Mechanik, Leipzig (1894).

11.4   Bell's Inequality

The previous sections inspire us to try hidden variable theories as a way to make the theory local and complete. That has been the subject of many attempts, but there has long since been a series of results, called "no-go" theorems, claiming the impossibility of local hidden variables—they would not be compatible with the empirical predictions of quantum mechanics. One of the cleverer "no-hidden variable" results is the Kochen–Specker theorem (1967): it places certain constraints on the kind of hidden variable that could exist.[4]

John Stewart Bell, working in CERN (Geneva) since 1960 and dividing his work between particle physics (90%) and foundations of physics (10%), was interested in "no-go" theorems. He understood, from the example of a theory of quantum mechanics worked out by David Bohm Bohm et al. (1952), that a hidden variable theory of quantum mechanics was in fact possible, but that this theory was nonlocal. In the end, he began to wonder whether it was possible to have a theory that was local and compatible with quantum mechanics . The answer turned out to be NO Bell (2004), and was the result of one of the most remarkable papers in the history of physics: "On the Einstein–Podolsky–Rosen Paradox" (1964). In Bell (1964), Bell pointed to some extraordinary features of entanglement. In particular, even if the two entangled components are far apart, they can influence each other. Bell demonstrated that there is no way to understand entanglement in a local framework.[5]

Let us add the (rather trivial) mathematics. Suppose that Z_α^i, $i = 1, 2$ denotes the predetermined outcome of the spin measurement in the EPR-B experiment for particle i along axis α. We have $Z_\alpha^i = \pm 1$ and $\alpha = a, b, c$ are three directions.

---

[4] It would in fact be important to explain how the Kochen–Specker theorem remains compatible with *certain* hidden variable theories. The point is that Kochen-Specker only shows the nonexistence of a noncontextual value map; see Bricmont et al. (2022), Bricmont (2016).

[5] Some authors like to add the adjective "real" and speak about "local realism." But what is the reason to introduce that difficult notion; after all, physics assumes at least some reality (independent of our existence) or what are we speaking about?

Clearly, if $Z_\alpha^1 = -Z_\alpha^2$ then the probabilities (interpreted in any way we like, possibly also just as proportions) satisfy

$$P[Z_a^1 \neq Z_b^2] + P[Z_b^1 \neq Z_c^2] + P[Z_c^1 \neq Z_a^2] \geq 1 . \tag{11.1}$$

That is the famous Bell inequality. Indeed, the three two-valued random variables cannot all disagree. However, when measuring over axes that are 60 degrees apart, each probability is just 1/4, which gives a contradiction. Actual measurements confirm the predictions of the quantum formalism, but we should emphasize that Bell's result is independent of quantum theory, so experimental results actually confirm a fact about Nature.

There is an extension of that Bell argument above, going by the name of Clauser–Horne–Shimony–Holt–Bell inequality, for short CHSH-Bell inequality (1969–1975). The aim here is a mathematically precise formulation for systems that have been produced by a common source and then are separated.

There are control parameters α_1 and α_2 that indicate which of several possible measurements are actually performed. These are chosen randomly and freely by the experimenters just before the measurement—the subscripts 1 and 2 refer to say left (Alice) and right (Bob) measurements. The outcomes are written as A_1 and A_2, with some statistical regularities described by probability distributions $P_{\alpha_1,\alpha_2}(A_1, A_2)$. There is no assumption of predetermined outcomes, and randomness can still occur during the measurement.

Let us turn to the premeasurement and preparation. All data available to both systems, e.g., from a common source, are codified in some random variable λ, so that we have the decomposition

$$P_{\alpha_1,\alpha_2}(A_1, A_2) = \int \mathbf{P}_{\alpha_1,\alpha_2}(A_1, A_2|\lambda) \, dP(\lambda) .$$

The probability $\mathbf{P}_{\alpha_1,\alpha_2}(A_1, A_2|\lambda)$ gives rise to expectations $\langle \cdot \mid \lambda \rangle_{\alpha_1,\alpha_2}$ for fixed λ and setting α_1 and α_2. This is given by the results of the measurement, and can be calculated in principle using the formalism of quantum theory. Assume that the probability distribution $P(\lambda)$ does not depend on α_1, α_2 (the settings of the measurement apparatus); this is called the *no conspiracy* condition: we assume that we are in sufficient control of what we want and what we decide to be the settings.[6] Next we turn to the locality assumption, which supposes that, when the two measurements are made very far apart,

$$\mathbf{P}_{\alpha_1,\alpha_2}(A_1, A_2|\lambda) = \mathbf{P}_{\alpha_1}(A_1|\lambda)\mathbf{P}_{\alpha_1}(A_2|\lambda) , \tag{11.2}$$

Or put another way, conditional probabilities factorize.

---

[6] Some authors make a bigger point of this. In fact, some Bell tests have been designed to avoid this "absence of free will" problem. Of course, we can never totally exclude the possibility of superdeterminism, just as we can never disprove solipsism.

Consider now the expected value of the product $A_1 A_2$, viz.,

$$C(\alpha_1, \alpha_2) = \int \langle A_1 A_2 | \lambda \rangle_{\alpha_1, \alpha_2} \, dP(\lambda) \, .$$

For a typical experiment (spin measurements on a pair of particles in the singlet state), the parameters $\alpha_{1,2}$ refer to directions, and they can be identified with unit vectors along some given orientation. Take four such directions $a, a', b,$ and c and suppose $A_1, A_2 = \pm 1$. Then we always have

$$\left| C(a, b) - C(a, c) \right| + \left| C(a', b) + C(a', c) \right| \leq 2 \, . \tag{11.3}$$

Proof By locality (11.2), $\langle A_1 A_2 | \lambda \rangle_{\alpha_1, \alpha_2} = \langle A_1 | \lambda \rangle_{\alpha_1} \langle A_2 | \lambda \rangle_{\alpha_2}$. Thus,

$$\left| C(a, b) - C(a, c) \right| + \left| C(a', b) + C(a', c) \right|$$
$$\leq \int \left[\left| \langle A_1 | \lambda \rangle_a \right| \left(\left| \langle A_2 | \lambda \rangle_b - \langle A_2 | \lambda \rangle_c \right| \right) + \left| \langle A_1 | \lambda \rangle_{a'} \right| \left(\left| \langle A_2 | \lambda \rangle_b + \langle A_2 | \lambda \rangle_c \right| \right) \right] dP(\lambda)$$
$$\leq \int \left[\left| \langle A_2 | \lambda \rangle_b - \langle A_2 | \lambda \rangle_c \right| + \left| \langle A_2 | \lambda \rangle_b + \langle A_2 | \lambda \rangle_c \right| \right] dP(\lambda) \, .$$

Finally, for real numbers $x, y \in [-1, 1]$, we always have $|x - y| + |x + y| \leq 2$.

The parameters a, b, c, a' can always be chosen so that the left-hand side of (11.3) becomes equal to $2\sqrt{2}$, as observed experimentally, and as quantum theory also predicts via the identity $C(\alpha, \beta) = -\alpha \cdot \beta$. (Choose b and c mutually orthogonal, with a' bisecting b and c, and a bisecting b and $-c$.) There is a clear contradiction—one speaks of Bell inequality violation. This establishes once again the incompatibility between locality (here represented in (11.2)) and empirical results (compatible with quantum calculations). In other words, there is indeed spooky action-at-a-distance.

11.5 Bell Test Experiments

The experimental verification of the violation of Bell's inequality (11.1) requires great care, and various steps with corresponding improvements have been taken over the years. There are in general two big challenges. First, one must have good detection. This means that the entangled particles must reach Alice and Bob to be measured there in very good statistical measure (detection loophole). Secondly, Alice and Bob must be far from each other, so that Alice can perform switching measurements in flight and no communication is possible between Alice and Bob during the motion (no light speed signaling loophole). Being farther apart helps as it gives a tiny bit more time for Alice to switch the measurement settings. Various ingenious solutions have been found for these problems, especially by the pioneers of the Bell tests.

The first three Bell test experiments performed by Alain Aspect in 1982 at Orsay used calcium cascade sources. The third test ensured that the choice between the settings on each side could be made during the flight of the photons (as originally suggested by Bell). It showed the inseparability of an entangled photon state, even if the photons were far apart, in fact at a "space-like" separation in the relativistic sense. This meant that no signal traveling at a velocity less than or equal to the velocity of light could connect the two measurements. The photons were separated by twelve meters at the time of the measurements, Aspect et al. (1981). In addition, it was possible to change the setting of the measuring polarizers during the twenty nanosecond flight of the photons between the source and the detector, Aspect et al. (1982), Aspect et al. (1982).

More recent experiments have used new sources, injecting entangled photons into two optical fibers. Detection efficiency has grown and the locality condition was enforced using variable-measurement apparatus. Separations of hundreds of meters and even more have been achieved, and it has even been possible to change the settings of the polarizers randomly during the propagation of the photons in the fibers. In 2015, a loophole-free Bell violation was reported using entangled diamond spins over 1.3 km and corroborated by experiments using entangled photon pairs, (Hensen et al. 2015; Shalm et al. 2015; Rosenfeld et al. 2017).

There have been quite a number of other Bell test experiments over the last ten years, including Ansmann et al. (2009), Giustina et al. (2013), Christensen et al. (2013), Larsson et al. (2014), Hensen et al. (2015), Shalm et al. (2015), Schmied et al. (2016), Handsteiner et al. (2017), Rosenfeld et al. (2017). Noteworthy examples include the BIG Bell Test Collaboration BIG Bell Test Collaboration (2018) was an international effort where human free will was used to close the "freedom-of-choice loophole." This was achieved by collecting random decisions from humans rather than using random number generators. Around 100 000 participants were recruited in order to provide sufficient input for the experiment to be statistically significant. This was citizen science for exploring the frontiers of Science.

Chapter 12
Future Experiments

The world, its people, and their occupations are changing fast. Yet, some things never change... . Just like in art and culture more generally, technological developments, often boosted by wars or big money, will remain an important force in the exploration of Nature. Science and physics in particular have truly meant a lot for the welfare and well-being of so many people and will continue to do that, Yet, all things considered, history is not necessarily there to follow or to be repeated. There are even intellectual and moral reasons to distance ourselves from some of the scientific and technological methods and products of the twentieth century. As we change and more people join the quest, physics will naturally take new directions,[1] hopefully less steered by (special or not) military operations or purely monetary goals, and less dominated solely by white male fancies. But surely (and optimistically), our interests and needs will shift, and a more mature understanding of the physics of the twentieth century will add to the fun.

And then, as we move deeper into the twenty-first century, we ask where the excitement is. Where are we heading? In a way, it is quite pointless to try to answer these questions, but think of it as preparing the day. While times are frantic, the future of physics and the excitement will be made by ourselves, friends, colleagues and by our future students. Often, it is individuals, performing tabletop experiments and asking new questions, who make the breakthrough, and there is less scope for fundamentally new physics in Big Science projects.

Therefore, the question to younger people is: for what physics—for what style and content of physics—would you be ready to leave your home country, your relatives, and your parents' house, bearing in mind that the action in physics is where experiments at the forefront of fundamental research are being conducted?

The test will be the power of quantitative prediction or explanation. Quantitative estimates or numerical results produced by experiments are essential for the whole enterprise of physics.

---

[1] Max Planck described "possible change" as follows: "A new scientific truth does not triumph by convincing its opponents and making them see the light, but rather because its opponents eventually die, and a new generation grows up that is familiar with it."

© The Author(s), under exclusive license to Springer Nature Switzerland AG 2023   143
C. Maes, *Facts of Matter and Light*,
https://doi.org/10.1007/978-3-031-33334-7_12

According to my views, aiming at quantitative investigations, that is at establishing relations between measurements of phenomena, should take first place in the experimental practice of physics. "By measurement to knowledge" I should like to write as a motto above the entrance to every physics laboratory.
— Heike Kamerlingh Onnes

A good experiment is "thinking and looking at the same time," but it also yields numbers. We have seen that in the Rutherford experiment with the determination of the size of the nucleus, in the Perrin experiment with the measurement of the Avogadro constant, in the measurement of the Lamb shift of course, and in the precise determination of the g-factor as part of its great success. Or take for instance the Stewart–Tolman effect, first demonstrated   in 1916 which, as written under the title "The electromotive force produced by the acceleration of metals", revealed the nature of electrical conductivity and measured the mass of the electron at the same time, Tolman and Stewart (1916). There are many other examples, not discussed here.

12.1 Around 2000

In 1905, Albert Einstein was preparing exams and writing publications while working as a clerk in the patent office in Bern. Six days a week with 8 hours' work, 8 hours' study, and 8 hours' sleep. His studies included the whole of physics as it was known around the year 1900. There were three main books on his desk. They coincided with the pillars of classical physics: electromagnetism, thermodynamics, and mechanics, see Fig. 12.1. The tensions between those three pillars form an important part of the context of the great performances discussed in the present book. After the tension came progress and that is the subject of many of your textbooks. Imagine yourself a student indeed, trying to master physics. What are the main books you are using and how do they relate to each other? What questions emerge? Where are the action and the excitement when you think about Nature today?

The twentieth century began with three major revolutions in physics: statistical mechanics, relativity, and quantum mechanics. The new insights are often described theoretically. But we should not make the mistake of thinking that theory alone had been towing the ship. The waves that shook physics and the clouds that gathered on its horizons around 1900 originated mainly from experimental evidence. Many of the great experiments of the twentieth century relate to the development of these changes. The previous chapters have dealt with some of them very explicitly, while others are related, and some are natural continuations. Let us repeat some key points.

Looking at Fig. 12.1, in the corners are the new domains of the twentieth century, unifying and solving problems at the interface of the classical pillars of physics. To begin with, statistical mechanics was not only able to reconcile thermodynamics and mechanics. It also managed to correct and extend thermodynamics, moving into the realm of fluctuations. Stochasticity entered fundamental considerations for connecting different levels of description and the power of large numbers was summarized via

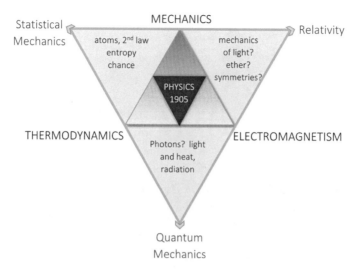

Fig. 12.1 The three revolutions at the beginning of the twentieth century released tensions between the pillars of classical physics around 1900

the concepts of entropic and fluctuation forces. The theory of emergent phenomena such as the arrow of time, phase transitions, and the appearance of chance was given a firm foundation. Moreover, the framework of statistical mechanics, the so-called Gibbs formalism, became the model for a whole range of major theoretical developments during the twentieth century. Much of quantum field theory followed the lines of Gibbs theory, and discussions on spontaneous symmetry breaking, universality, and on renormalization were inspired by it. It parallelled the development of condensed matter physics. In the period 1960–1990, there was a fruitful exchange of ideas and constant interaction between high-energy physics and condensed-matter communities. Today, computer science, from optimization problems to machine learning and data mining and all the way to artificial intelligence, are based on inference schemes pioneered in statistical mechanics.

Secondly, relativity was able to unify EMO with mechanics. Newtonian mechanics had to be upgraded but in fact, it began by truly unifying electricity, magnetism, and optics, in ways that went far beyond Maxwell's theory. Naturally, this affected the very foundations of our view of spacetime. In particular, its later development in the general theory of relativity provided what has become the classical foundation of our concept of spacetime. Here, gravity becomes a field theory, and the basis for our standard model of cosmology. Together with quantum mechanics, special relativity was built into quantum electrodynamics and finally into the Standard Model of elementary particles.

Thirdly, quantum mechanics is a completely new kind of mechanics, and a generalization of Newton's program. It resolved various phenomenological mysteries and relaxed the tensions that had arisen in physics regarding the interaction between light and matter. It enabled us to look into the subatomic world and analyzed the structure

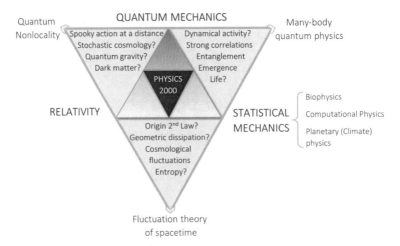

Fig. 12.2 What are the revolutions in the twenty-first century that will release the tensions between the pillars of physics as they stood around 2000?

of matter and radiation. In short, quantum physics was a complete game-changer, enabling quantum statistical mechanics and condensed matter theory to reach as yet unimagined realms of thought.

The new "triangle" of physics that came into being at the beginning of the twentieth century had relativity, quantum mechanics, and statistical mechanics at its corners (see Fig. 12.1). Again, many intersections or overlaps have arisen, with the usual tensions and the challenge of making them compatible. Today, it is hard to support the whole construction we now call physics on those same three corner pillars that were erected about 100 years ago. We may expect the new developments to allow more revolutions, and give rise to new directions and fields, perhaps as in Fig. 12.2.

Now it is the older triangle in Fig. 12.2, statistical mechanics, relativity, and quantum mechanics, that has to make place for newer corners of investigation. Let us try to consider the newest tensions between them and the challenges they will bring.

12.1.1 The Statistical Mechanics of Geometry

Experimental geometry..., the combination does not appear to exist, while, all the same, geometry is one of the oldest natural sciences. Still, today various research programs in soft and hard condensed matter go into that direction. We then mostly speak, not so much of geometry but of, topology and metric structure where available. Subjects such as topological insulators or Hall conductivity (edge states) and cold atoms touch these themes in hard condensed matter, and the study of the influence of curvature, twisting and bending of structures on mechanics, transport and (biological) functioning are by now standard in soft condensed matter. There is an interesting development there and even beyond, how the geometric degrees of freedom have become active and equal partners in the interaction with matter. There, we would be helped by experimentally probing the fluctuations in geometry.

The relations between matter and heat, between heat and light, and between matter and light have been a recurrent theme of twentieth century physics. Still, slowly in many of these studies, the substrate of spacetime has become a dynamical counterpart. Architecture, metric structure, topology and geometry are no longer the passive background. We see that in the physics of disordered systems and in duality relations, where geometry (e.g. in the form of cluster representations) and matter start to play equivalent roles. Studies of (random) networks or graphs, whatever is their origin, have become relevant experimental physics. We have of course and since long a great example: general relativity equates properties of spacetime and matter. But we need more. With some exaggeration, but only slightly so, there is no statistical mechanics of general relativity. Trivially, there are ways to write thermodynamics covariantly, and much of the well-written history of our universe concerns heat and radiation. The magnetohydrodynamics of plasma, energy conversion, and structure formation are also central topics in statistical mechanics. Sure enough, the thermal history of our universe has already been written, even as early as 1934 by Richard Tolman in his classic book *Relativity, Thermodynamics and Cosmology*, which extends thermodynamics to special and general relativity. Applications of relativistic mechanics and relativistic thermodynamics are obviously needed in the construction and interpretation of cosmological models.

However, there is no fluctuation theory for the (pseudo-)Riemann geometry used in the description of our cosmos. We have no geometric entropy or geometric dissipation, and we have little idea of nonequilibrium effects, be they transient or steady, that (would or could) drive the universe as a whole. In fact, the statistical mechanics of Newtonian gravitation is already complicated and suffers from many difficulties. One immediate source of problems is the long-range and (universally) attractive nature of Newtonian gravity. Instabilities, collapse, and structure formation give rise to hard problems. Some of the problems could perhaps be circumvented or even eliminated by introducing geometry and its field dynamics to replace Newton's force, but, as I mentioned, that has not yet been achieved. Some of us even hope to understand the emergence of spacetime and understand gravity as an entropic force, but we are still a long way from achieving that. Statistical cosmology, not in the sense of descriptive statistics but as a fluctuation theory for spacetime, even in a classical form, is one thing we are looking forward to. Statistical Riemannian geometry, statistical cosmology, or statistical relativity; the name must still be given to a newer and better unification of statistical reasoning and fluctuation theory with the general theory of relativity.

Fluctuations are corrections to typical behavior and the latter requires specifying the level of description. Bridging different levels and transferring more microscopic ingredients to more macroscopic behavior, is what we have entropy for. It is not clear what we should mean by the entropy of a geometry. However, coming to terms with "entropy" is not everything; the notion of energy is also somewhat problematic in general relativity. There appears to be no such thing as the density of gravitational energy. Moreover, spacetime can do (positive or negative) work on matter. In fact, therein may lie a clue to dealing with the thermal properties of geometries.

Back to experiments. Obviously, geometry and its characteristics such as curvature are not only relevant to gravity. Quite important biological functioning is steered or influenced by mechanical deformations, e.g., changes in architecture or substrate. We can have in mind mechanical transduction, which is the transformation of (cellular) deformation into an electrochemical response. This is essential to the survival of cells and even higher organisms. Thinking for instance about stem cell behavior in development and regeneration, the imposed geometry and changes in elastic or mechanical forces are decisive. Clearly, that mechanics tends to be on small (mesoscopic) scales and hence subject to chemical and thermal fluctuations. That includes studies in micro-rheology and nonequilibrium elasticity, where fluctuating small-scale topology opens new avenues in mathematics and in applied sciences alike. Another example is the use of techniques from general relativity for a better understanding of heart rhythm disorders.

Broadening the idea, self-organized, responsive, and adaptive geometrical structures like those in morphing and active matter, are the subject of statistical mechanics. Moreover, topological characterizations and classifications play a growing role in soft and hard condensed-matter physics. Experimentally, the study of self-assembly, fluctuating geometries, and biomimetic and biohybrid structures are also of growing importance.

12.1.2 Relativity Versus Quantum Mechanics

Special relativity and quantum mechanics meet each other successfully in quantum electrodynamics and in the unification of the electroweak and strong interactions. It is also possible to describe quantum corrections to Newton's law of gravity. But what about quantizing spacetime and obtaining quantum corrections to Einstein's field equations? In that view, gravity, as described in general relativity, is merely an effective field theory at low energies and macroscopic distances.

There is no quantum extension of general relativity, and some would say there is no need. Naturally, such a theory would focus on specific problems of cosmology to discover quantum features, but they are far from observable. One is the physics of black holes, where a thermodynamic analogue has predicted that black holes have entropy, which is of course not surprising.[2] The statistical mechanics of black holes would need ingredients from both quantum theory and general relativity. Another more general subject is to understand the role of entanglement in cosmological history.

---

[2] One approach is to try to unravel the structure of black holes, probing (theoretically) internal degrees of freedom by counting them. Another less trodden path is to think about thermal properties and extend the thermodynamics of black holes to astrophysical reality, for example, taking into account a positive cosmological constant. Obviously, that must refer to statistical geometry again, as discussed in the last section.

It is connected with this other vast field of research, now supplied with increasing observational information, which is the physics of the (so-called) early universe. There again, we hope to see an interesting interplay between quantum physics and general relativity.

Quantum gravity is also connecting with themes in condensed matter physics under the general denominator of holography. Indeed, entanglement and spacetime structure regularly meet in puzzles arising from the many-body physics of strongly correlated condensed matter systems. A big experimental challenge is to see how the theoretically developed ideas of holography can be realized experimentally. This clearly touches upon the next topic. A more specific realization is in the Sachdev-Ye-Kitaev model, an exactly solvable model that relates both to the physics of strongly correlated materials and to the ideas of holography and quantum gravity. Interestingly, quantum simulation appears to be possible, Luo et al. (2019).

And then there is the big puzzle how the nonlocality of quantum mechanics is to be reconciled with Lorentz invariance. Perhaps the latter is only emergent, and the nonexistence of absolute time is the more fundamental property. It is not clear here what experiments will contribute to a solution.

12.1.3 Quantum Statistical Mechanics

The previous section hints at new developments in quantum many-body physics in connection with strongly correlated entities or strongly coupled systems. There are of course the older themes of quantum statistical mechanics, such as the quantum corrections to specific heats and the phenomena of superconductivity and superfluidity. There are, however, many newer themes such as high-temperature superconductivity, topological effects in condensed matter physics, and the search for anyons, where nonperturbative effects are essential. We refer to the books Sachdev (2011), Carr (2010) for an overview.

What we would stress here, however, is the appearance of a new field, somewhere between quantum theory and statistical mechanics. Perhaps for lack of a better term, this is often called quantum information theory. It deals with a large number of problems, but it sprang mainly from the possibility of gaining control over individual quantum particles. The manipulation of small systems and the ability to control and measure their quantum statistical properties is a theme that first came up in the 1980s. An important application here is quantum optics and combined with condensed matter physics, it is up for the second quantum revolution. It is also giving new boosts to metrology, and to computational physics where it provides novel quantum simulation techniques. Although this whole new field goes under the name of quantum information theory, we could also call it (and would do better to call it) quantum many-body physics. Entanglement is a key player here.

> I would not call that *one* but rather *the* characteristic trait of quantum mechanics, the one that enforces its entire departure from classical lines of thought.
> E. Schrödinger, In: Discussion of probability relations between separated systems, 1935.

Strangely, few decades ago, there was little or no mention of entanglement in the books on quantum field theory. What is known about entanglement now plays an increasing role in statistical mechanics. It refers of course to a key feature of quantum many-body physics. Previous approximations, variational estimates, or perturbative treatments too often used a basis of non-entangled quantum states. This is changing, using the methodology of matrix states and tensor networks. There, entanglement is cherished and appreciated from the outset. Quantum aspects of relaxation behavior or the absence thereof (localization) may relate to the role that entanglement plays there.

Other notions such as quantum dissipation and quantum dynamical activity are important and intriguing topics of statistical mechanics, where we move into the nonequilibrium world at quantum scales of energy.

Here as well, experiments are likely to demolish dogmatic thinking and modify textbook quantum physics. One line of development that will most probably remain active is the search for quantum trajectories, where particles move nonclassically through space and in doing so are guided by a wave function. This simply is the best way to make sense of the quantum world; see e.g. Bricmont (2016).

Experimental developments are now in full swing. Quantum optics forms the substrate for the various ways the second quantum revolution is being realized. Modern developments in quantum optics have led to new techniques, often based on controlling and exploiting atom–photon interactions. Work by experimentalists like Ching-Wu Chu, Claude Cohen-Tannoudji and William Phillips has established the use of laser cooling and trapping of atoms. Another new invention is the cold atom clock, which has already allowed time measurement to reach a level of one-second accuracy in thirty million years.

Concerning quantum technology for quantum computing, sensing, and transport, note Bachman's words: *you ain't seen nothing yet*. The manipulation of single atoms or single quantum states and the exploitation of many-body quantum entanglement are the main breakthroughs of recent days.

12.2 Into the Next Hundred Years

We may speculate that three central problems, gravity and its fluctuating geometry, the physics of life, its origin, and biological functioning, and the impossible emergence of consciousness and related mind–body problem, will be among the central themes of physics for the next hundred years, accompanied by exceptional bio-and quantum sensing methods, and as yet unseen levels of numerical and simulation capacity, as the main tools. In the short run, the technology resulting from many-body quantum physics in general, and from (long range) entanglement more specifically, will most likely surprise us even more.

However, the problems or unification attempts mentioned above and emphasized through Fig. 12.2, statistical cosmology or geometry, quantum gravity or holographic physics, and quantum information theory or many-body physics, are far from accounting for all today's physics challenges. Let us roam in some other directions as well.

12.2.1 Computational and Neuro(nal) Physics

We are constantly being assisted by machines to probe and observe Nature. Machines, engines, and electronics make the instruments for experimenting, including, in a growing and combined way, an array of information technologies. The word "machine" gets a somewhat different meaning in the area of machine learning. It stresses the algorithmic side of "learning" from data. Machine learning and deep learning are certainly on the rise in diagnosing or monitoring physical systems. But it remains to be seen how, combined with the newest algorithms for data mining, all that will answer interesting physics questions. Perhaps it will just change the questions.

Computers obviously play an important role in physics, if only to compute and sort data. Computational physics and AI instrumentation are here to stay and to grow with us. Every form of physics education will invest in those skills; programming, simulation, and visualization of physics and mathematics are the required expertise. Perhaps one day, zombies, computers, or machine software will do physics, even experiments. On the other hand, some information technology results directly from monitoring physics experiments.[3] Interestingly, many great developments have been heavily influenced by work in statistical mechanics. Ways to deal with highly inhomogeneous systems and multiscale dynamics are indeed traditional themes there; consider the 2021 Nobel prize *for the discovery of the interplay of disorder and fluctuations in physical systems from atomic to planetary scales*, awarded to Giorgio Parisi, famous for his work on spin glasses and related statistical mechanical models relevant to optimization theory. Earlier, highly influential models of neural networks, such as the Hopfield model, were inspired by statistical mechanical models such as

---

[3] The internet being the prime example.

the Ising model. Elisabeth Gardner (1957–1988) in particular used the theory of spin glasses for essential contributions to the study of neural networks; her contributions to learning, optimal storage, and related phase transitions were groundbreaking there.

On the other hand and not completely unrelated, the physics of the brain and information processing will be one of the big themes in the coming years. It has a much longer history, e.g., with pioneers such as Ada Byron (1815–1852), the countess of Lovelace and the world's first computer programmer. She set out to create a mathematical model and calculus of the nervous system, to understand how the brain gives rise to thoughts and nerves to feelings. This theme is even broader than Artificial Intelligence, and also moves in other directions. Even without considering questions relating to the emergence of consciousness, there are many aspects of our own brain's functioning we do not understand. Clearly, the brain is active, producing a great deal of heat, and it is plastic (at least, it was in the beginning), performing many complicated tasks such as memory, representation, sensory adaptation, etc. It needs no longer be the sole domain of neurobiology. Physics will reach out to it and, as in the times of Helmholtz and other pioneers of physiology, physics will also learn from it. Here as well, as in general relativity, we have an interesting interaction between geometry and motion, and in particular between network features and physical transmission and storage. How indeed could we understand the structure of stimulus–response functions without considering the nonequilibrium statistical mechanics of neuronal networks and physical interactions together?

Perhaps there is a major challenge for physics here, to discover its own limits.

We can count on it
when we're sure of nothing
and curious about everything.

Among the material objects
it favors clocks with pendulums
and mirrors, which keep on working
even when no one is looking.

It won't say where it comes from
or when it's taking off again,
though it's clearly expecting such questions.

We need it
but apparently
it needs us
for some reason too.

— From the poem "A Few Words On The Soul," by Wislawa Szymborska (1923–2012), translated from the Polish by Stanislaw Baranczak and Clare Cavanagh.

Indeed, many of us physicists are not physicalists, meaning that we believe that physics, for all its impersonal and universal objectivity, cannot be sufficient to understand the idea of a mind or of a conscious and feeling subject. That would not exclude its going a long way toward understanding and mimicking brain functioning, and even outrunning it in various tasks under some heading of "Artificial Intelligence." Before we get there, however, many "simple" physics questions are still unanswered, such as how to catch the essence of self-organization, viz., the organization of the "self", and how to understand the stability of neuronal networks, the plasticity of brains, and life itself.

12.2.2 Physics of Life

When we have found how the nucleus of atoms is built up we shall have found the greatest secret of all — except life. We shall have found the basis of everything — of the earth we walk on, of the air we breathe, of the sunshine, of our physical body itself, of everything in the world, however great or however small — except life.
— Ernest Rutherford.

Around the year 2000, biophysics took off seriously, and attention turned massively to the physics of cell biology on the one hand and ecology on the other. The number of poorly understood themes in the physics of life is enormous, but clearly, the mysteries of life, of evolution, and of consciousness are at least as exciting as the origin of our material universe.

There are at present fundamental problems in theoretical physics awaiting solution, e.g., the relativistic formulation of quantum mechanics and the nature of atomic nuclei (to be followed by more difficult ones such as the problem of life), the solution of which problems will presumably require a more drastic revision of our fundamental concepts than any that have gone before.
— Paul Dirac, Proc. R. Soc. London, Ser. A **133**, 60 (1931).

Biophysics and the physics of life and for life are indeed attracting increasing interest. That is a good thing. The time is long gone when it was thought evident that physics was only about breaking things up into smaller and smaller pieces. The time has come to look at global structures and functioning, from nanoscale biology to ecology. In a sense, fundamental physics is coming home here, as it finally seeks to understand the nature of Nature. Perhaps the physical universe is more alive than we have yet imagined.

The four chemical elements, C, H, O, and N, make up over 99% of all organic tissue on Earth. The cell is the basic unit, open and yet largely independent, alone and in interaction, metabolic and stable. A cell contains some 100 trillion atoms, however. Clearly, understanding the physics of a complex system the size of a typical cell is dauntingly difficult. Useful approximations and new models are badly needed; ones that identify and take into account what is essential. It remains to be seen what the main physical principles will be, what is most important among them, and what are just details. The history of statistical mechanics can provide us with clues about how to proceed. (Active) phase segregation is probably essential in the understanding of cell division. One exciting theme here is the combination of architecture, in the sense of mechanical stresses and forces, and biochemical functioning. We seem to need the physics of bioelasticity to understand mechanical transduction and the role of stretching, curving, folding, and twisting substrates and macromolecules.

Chemical reactions and biological processes often run far from equilibrium. There is no formalism here comparable to that of Gibbs for the description of systems out of equilibrium. The phenomenology also gets so much richer and we really need experiments to probe the kinetic effects. Fluctuation theory is not in fact sufficiently well understood for active processes, and nonequilibrium response theory, the main model for experimental science, still lacks applications in the life sciences.

Experimentally, optical tweezers, artificial motors, femtosecond imaging, and data-mining techniques make it increasingly possible to monitor biological functioning in action. For example, optical tweezers can grab and steer particles, such as bacteria, viruses, and other living cells with their laser beam fingers, using the radiation pressure of light. The work on cooling and trapping of atoms referred to in Sects. 10.4–10.6 was empowered by such optical micro-manipulations. Artificial and molecular motors are synthetic or natural devices that consume energy in one form (powered chemically or optically) and convert it into motion. Rotary molecular motors driven by light were first created in 1999 in the laboratory of Ben Feringa. This allows us to steer and unravel mesoscopic trajectories to an as-yet unheard-of precision. We are thus faced with so many biodata that physics really will be needed to identify the crucial experiments and the relevant theory.

On the other hand, when we contemplate physics *for* life, there is the growing and crucially important art of biosensing. Soon, we will be able to "see" the manufacture and working of drugs, and online and fast diagnosis is already within reach. Clearly, medical applications will soon follow these experimental achievements. Donna Strickland won the 2018 Nobel prize with Arthur Ashkin (1922–2020), father of optical tweezers, and Gérard Mourou, her doctoral adviser, for her work in ultrafast and ultrabright laser physics, widely used now in the study of biological systems. In particular, her work led to the shortest and most intense laser pulses ever created, also known as chirped pulse amplification, which has found uses in therapies targeting cancer and in the millions of corrective laser eye surgeries that are performed each year. Anne L'Huillier, cowinner of the 2022 Wolf Prize, has also been a key scientist in founding the new field of attosecond research.

12.2.3 Many-Body Nonequilibrium Physics

For some time now, an interesting wave of soft condensed matter physics has been running through science departments around the world, where emphasis more often goes on nonequilibrium physics. The same is happening now for solid state physics (hard condensed matter) and optics. In general, the dynamics of systems influenced by fluctuations far from equilibrium or determined by strongly nonlinear interactions is rapidly gaining interest from all sides of physics.

In general and in principle, we can distinguish between two types of structures or processes in Nature. There are those "in detailed balance," for example in crystal formation or in the evaporation of water, and there are those that are "out of balance," producing or even maintaining currents of particles, energy, and momentum, while being stable and even stationary on the relevant timescales. Here, we might think of ourselves, but there are many much simpler examples. Irreversible dissipation is often the consequence of being open and, combined with "driving", the system settles in a steady nonequilibrium state.

Nonequilibria are indeed everywhere, from cosmological scales in early universe plasma physics to nanometer transport in quantum devices. Experiments on nonequilibrium systems have been increasing fast, e.g., in colloidal physics, biophysics, and the study of complex fluids and driven optical lattices. Still, there are many fundamental questions that wait to be linked with experiment, e.g., regarding low-temperature thermal properties, characterizations of susceptibilities and the corresponding response theory, or again critical phenomena and phase transitions in driven and active matter. Theory is still tentative there and often restricted to gases (dilute systems). We are building on a many-body framework that would give exciting new additions to the theory of localization, e.g., in glasses, and to questions of self-organization. These remain subjects to be explored experimentally in the coming decades.

On the more applied side, the study of morphing matter has emerged as a cross-disciplinary field. The idea is to create responsive, active, and adaptive material systems, e.g., as seen in plants and other growing organisms. Fundamental challenges are found here too, in the relationship between inert and living matter. We speak of biomimetic and biohybrid systems.

As a specific case, we also mention the problem of many-body chaos. Turbulence has been called the last great unsolved problem of classical physics. And here again, many uncertainties exist even in the phenomenology. Moreover, turbulent drag, which we are sure we see, remains unexplained theoretically from a more microscopic point of view. Similarly, the phase transition from a laminar to a turbulent regime is part of a larger set of transitions in which the phenomenology soon gets too broad and uncertain to even speak of one theory.

The most promising approach would seem to be to consider turbulence as a problem in nonequilibrium statistical mechanics. The system is extremely intricate, involving many scales, and rich but complicated time-dependent behavior can follow. Consider a pot of water that is heated at the bottom and cooled down at the top. It is out of balance. If the temperature difference increases, convection currents will arise. This kind of flow consists of water molecules that enter large cylindrical "rolls" and regulate heat transport from bottom to top. Such convection rolls are robust. They can be found outside the kitchen too, e.g., currents in planetary atmospheres (such as our own) and beneath stellar photospheres (for example, in the Sun). If the difference in temperature is even greater, more complex structures can arise, and flow patterns can complicate considerably. We then reach the regime of turbulence, where even very small obstacles or inhomogeneities can be blown up and manifest themselves in different flow profiles. Similar turbulent flows can also be obtained by large pressure differences or by external driving forces. Such chaotic macroscopic systems have a very complex, critical, and to some extent even law-free phenomenology. Yet, certain properties are reproducible, except that no systematic phenomenology has yet been found. Experiments on turbulence and far-from-equilibrium systems are sure to see exciting developments.

Topics of growing interest also include ultrafast light pulses and spin dynamics. To miniaturize a quantum optics table on an integrated chip meets with fundamental problems of nonequilibrium physics. Already now, nonlinear integrated photonic circuits are becoming increasingly accessible. Controlling spin transport is another challenge, and in particular how to understand there extended fluctuation-dissipation relations. Related to that is the study of spontaneous magnetization fluctuations under nonequilibrium conditions. Moving then to quantum materials, there is the rich (but very open) phase diagram of transition metal oxides. There is the interplay of dynamical processes involving electrons, phonons and magnons. Those phase diagrams will be more and better explored by light-based methods, opening the fields of quantum nonequilibrium spectroscopy. Surely, we expect that the statistical features of light will get coupled with electronic fluctuations and *vice versa*. The experimental challenge is to quantify the role of fluctuations in nonlinear light–matter interactions. The thermal consequences remain barely explored as well.

12.2.4 Climate and Planetary Sciences

Climate science is very much part of physics. The Nobel Prize in Physics 2021 was awarded "for groundbreaking contributions to our understanding of complex systems," with one half jointly "for the physical modeling of Earth's climate, quantifying variability and reliably predicting global warming." The winners were Syukuro Manabe and Klaus Hasselmann, the former for pioneering the first modern climate models, including a precise treatment of the greenhouse effect and predictions, and the latter for developing stochastic climate models (and the study of weather fluctuations), including the identification of anthropogenic effects.

Geophysics, fluid dynamics, meteorology, oceanography, and so on, are all part of climate modeling, and let nobody tell you that this is not fundamental physics. For one thing, unraveling the history and the future of the Earth and its place in the Solar System and in the universe at large requires basic scientific culture and competence. All this requires the superior science of thermodynamics[4] to understand the physics, with its (as yet unsolved) extensions to irreversible phenomena. We have to deal with beautiful and fundamental questions about the origin and stability of structure, self-organization, current fluctuations, and response in complex systems, working out how to meaningfully relate multiple length and time scales within them and cleverly compute and simulate them.

---

[4] Einstein would call thermodynamics a superior field of science, being "the only physical theory of universal content of which I am convinced that, within the framework of the applicability of its basic concepts, it will never be overthrown."

But that is not all. In our times, we have urgent duties toward future generations. Some of the problems that must be solved concern no less than the future well-being of the human population. Global warming, declining biodiversity, and the difficult heritage of polluted land and waters hang like dark clouds over our schools and universities, which themselves struggle to remain relevant and *free* educational and emancipatory institutes.

The same holds for the even broader topic of planetary physics and astrobiology. Our survival is not the only motivation; we want to know just how typical life and the Earth actually are in the universe, to see ourselves in the light of the bigger picture that science and physics in particular may offer us.

> Science is the titanic endeavor of the human intellect to break out of its cosmic isolation through understanding.
> — Prof Nummedal, In: *Beyond Sleep* by Willem Frederik Hermans.

One might think that I refer here to space exploration or to various kinds of astronomical missions, or to Big Data and the enormous computational efforts that are often needed. All that may of course be crucial, but experiments should and will contribute a great deal as well. They often combine geochemical and fluid-dynamical approaches. This is clear for advancing technology in the context of sustainable energy or thermodynamic efficiency. And once again, it would be a grave mistake to conclude that there is no new physics here. Things like photosynthesis and cellular respiration, convection-driven dynamo effects, phase separation under active forces, or stress-induced chemical physics are all related to deeper questions of active processes that may very well be basic to understanding our universe but are not yet incorporated in any fundamental theory of Nature.

Indeed, many fundamental problems lie in understanding the role of nonequilibrium features and self-organization. As an example, capitulation to anthropic principles, fine-tuning arguments, and related speculations about multiverses are partially caused by ignoring the possibility of self-assembly and of the formation of structures that would be unstable under equilibrium conditions. At any rate, the relevant questions are still in the making, and so are still far from receiving a definite direction, let alone an answer. As a matter of fact, rethinking the questions of Nature remains a fundamental duty as well.

References

Allen, J., & Misener, D. (1938). Flow of liquid helium II. *Nature, 141*, 75.

Anderson, P. (1972). More is different. *Science, 177*(4047), 393–396.

Anderson, M. H., Ensher, J. R., Matthews, M. R., Wieman, C. E., & Cornell, E. A. (1995). Observation of Bose-Einstein condensation in a dilute atomic vapor. *Science, 269*, 198–201.

Andrews, T. (1869). The Bakerian Lecture: On the continuity of the gaseous and liquid states of matter. *Philosophical Transactions of the Royal Society of London, 159*, 575–590.

Ansmann, M., et al. (2009). Overcoming the detection loophole in solid state. Violation of Bell's inequality in Josephson phase qubits. *Nature, 461*, 504–506.

Aspect, A., Dalibard, J., & Roger, G. (1982). Experimental yest of Bell's inequalities using time-varying analyzers. *Physical Review Letters, 49*(25), 1804–1807.

Aspect, A., Grangier, P., & Roger, G. (1981). Experimental tests of realistic local theories via Bell's theorem. *Physical Review Letters, 47*(7), 460–463.

Aspect, A., Grangier, P., & Roger, G. (1982). Experimental realization of Einstein-Podolsky-Rosen-Bohm gedankenexperiment: A new violation of Bell's inequalities. *Physical Review Letters, 49*(2), 91–94.

Balmer, J. J. (1885). Notiz uber die Spectrallinien des Wasserstoffs. *Annalen der Physik, 261*(5), 80–87.

Batchelor, G. K. (1972). Sedimentation in a dilute dispersion of spheres. *Journal of Fluid Mechanics, 52*(2), 245–268.

Batchelor, G. K., & Green, J. T. (1972). The determination of the bulk stress in suspension of spherical particles to order c^2. *Journal of Fluid Mechanics, 56*, 401–427.

Bednorz, J. G., & Mueller, K. A. (1986). Possible high T_c superconductivity in the Ba-La-Cu-O system. *Z. Physik B Condensed Matter, 64*, 189–193.

Bell, J. S. (1964). On the Einstein-Podolsky-Rosen paradox. *Physics, 1*, 195–290.

Bell, J. S. (2004). *Speakable and unspeakable in quantum mechanics* (2nd ed.). Cambridge: Cambridge University Press.

Bethe, H. A. (1947). The electromagnetic shift of energy levels. *Physical Review, 72*, 339.

Bohm, D. (1951). *Qantum theory*. Englewood Cliffs, N.J.: Prentice-Hall.

Bohm, D. (1952). A suggested interpretation of the quantum theory in terms of "hidden variablesâŁž, Parts 1 and 2. *Physical Review, 89*, 166–193.

Bohm, D. (1953). Comments on an article of Takabayasi concerning the formulation of quantum mechanics with classical pictures. *Progress of Theoretical and Physics, 9*, 273–287.

© The Editor(s) (if applicable) and The Author(s), under exclusive license to Springer Nature Switzerland AG 2023
C. Maes, *Facts of Matter and Light*,
https://doi.org/10.1007/978-3-031-33334-7

Bragg, W. H., & Bragg, W. L. (1913). The reflexion of X-rays by crystals". *Proceedings of the Royal Society of London, Series A: Mathematical and Physical Sciences, 8*(605), 428–438.

Bricmont, J. (2016). *Making sense of quantum mechanics.* Springer.

Bricmont, J. (2022). *Making sense of statistical mechanics.* Springer.

Bricmont, J., Goldstein, S., & Hemmick, D. (2022). From EPR-Schrödinger paradox to nonlocality based on perfect correlations. *Foundations of Physics, 52,* 53.

Callen, H. B., & Welton, T. A. (1951). Irreversibility and generalized noise. *Physical Review, 83,* 34.

Carr, L. D. (2010). *Understanding quantum phase transitions.* CRC Press.

Casimir, H. B. G. (1948). On the attraction between two perfectly conducting plates. *Proceedings of the Koninklijke Nederlandse Akademie van Wetenschappen B, 51,* 793–795.

Casimir, H. B. G., & Polder, D. (1948). The influence of retardation on the London-van der Waals forces. *Physical Review, 73,* 360–372.

Chandrasekhar, S. (1949). Brownian motion, dynamical friction, and stellar dynamics. *Reviews of Modern Physics, 21,* 383.

Christensen, B. G., et al. (2013). *Physical Review Letters, 111,* 130406.

Compton, A. H. (1921). *The magnetic electron. Franklin Institute Journal, 192*(2), 145–155.

Compton, A. H. (1923). A quantum theory of the scattering of X-rays by light elements. *Physical Review, 21*(5), 483–502.

David, K. B., Mewes, M. O., Andrews, M. R., van Druten, N. J., Durfee, D. S., Kurn, D. M., & Ketterle, W. (1995). Bose-Einstein condensation in a gas of sodium atoms. *Physical Review Letters, 75,* 3969–3973.

Davisson, C., & Germer, L. H. (1927). The scattering of electrons by a single crystal of nickel. *Nature, 119*(2998), 558.

Davisson, C., & Germer, L. H. (1928). Reflection of electrons by a crystal of nickel. *Proceedings of the National academy of Sciences of the United States of America, 14*(4), 317–322.

Davisson, C., & Kunsman, C. H. (1922). The Scattering of Electrons by Nickel. *Physical Review, 19,* 253.

Davisson, C., & Kunsman, C. H. (1923). The scattering of low speed electrons by platinum and magnesium. *Physical Review, 22,* 242.

de Broglie, L. (1924). *Recherches sur la théorie des quanta.* Paris, 1924; *Annales de Physique,10*(3), 22–128.

Dijksterhuis, E. J. (1950). Mechanisering van het wereldbeeld (English translation: The Mechanization of the World Picture, published in 1961, also as The Mechanization of the World Picture: Pythagoras to Newton).

Duerinckx, M., & Gloria, A. On Einstein's effective viscosity formula. arXiv:2008.03837v3 [math.AP]

Duminil-Copin, H., 100 years of the (critical) Ising model on the hypercubic lattice. arXiv:2208.00864v1 [math.PR]

Dürr, D., Golstein, S., Tumulka, R., & Zanghì, N. (2003). Trajectories and particle creation and annihilation in quantum field theory. *Journal of Physics A: Mathematical and General, 36,* 4143–4149.

Einstein, A. (1905). Über die von molekülarkinetischen Theorie der Wärme geforderte Bewegung von in ruhenden Flüssigkeiten suspendierter Teilchen. *Annalen der Physik, 17,* 549–560.

Einstein, A. (1905). Uber einen die Erzeugung und Verwandlung des Lichtes bettrefenden heuristischen Gesichtspunkt. *Annalen der Physik, 17,* 132.

Einstein, A. (1906). Zur Theorie der Lichterzeugung und Lichtabsorption. *Annalen der Physik, 20,* 199.

Einstein, A. (1910). The theory of opalescence of homogeneous fluids and liquid mixtures near the critical state. *Annalen der Physik, 33,* 1275–1298.

Einstein, A. (1925). Quantentheorie des einatomigen idealen Gases. *Sitzungsber. Preussischen Akad. Wiss., 1,* 3–30.

Einstein, A., Podolsky, B., & Rosen, N. (1935). Can quantum-mechanical description of physical reality be considered complete? *Physical Review, 47*, 777–780.

Faraday, M. (1839). I. Experimental researches in electricity—fifteenth series.*Philosophical Transactions of the Royal Society of London*, (129), 1–12.

Feynman, R. (1995). Lectures on Gravitation (Ed. B Hatfield) Redwood City, Calif.

Freedman, S. J., & Clauser, J. F. (1972). Experimental test of local hidden-variable theories. *Physical Review Letters, 28*(14), 938.

Freire Oliveira, G. E., Maes, C., & Meerts, K. (2021). On the derivation of the Kompaneets equation. *Astroparticle Physics, 133*, 102644.

Freire Oliveira, G. E., Maes, C., & Meerts, K. (2022). Photon frequency diffusion process. *Journal of Statistical Physics, 189*, 4.

Friedrich, W., Knipping, P., & von Laue, M. T. F. (1912). Interferenz-Erscheinungen bei Röntgenstrahlen. Sitzungsberichte der Mathematischen-Physikalischen Klasse der Königlich Bayerischen Akademie der Wissenschaften zu München 303–322.

Gea-Banacloche, J. (2002). Splitting the wave function of a particle in a box. *American Journal of Physics, 70*, 307–312.

Geiger, H. (1910). The scattering of the α-particles by matter. *Proceedings of the Royal Society of London. Series A, Containing Papers of a Mathematical and Physical Character,83*(565), 492–504.

Geiger, H., & Marsden, E. (1913). LXI. The laws of deflexion of α particles through large angles. *The London, Edinburgh, and Dublin Philosophical Magazine and Journal of Science,25*(148), 604–623.

Geiger, H., & Marsden, E. (1909). On a diffuse reflection of the α-particles. *Proceedings of the Royal Society A: Mathematical, Physical and Engineering Sciences, 82*(557), 495–500.

Georgii, H.-O., Häggström, O., & Maes, C. (2021). The random geometry of equilibrium phases. In C. Domb & J. L. Lebowitz (Eds.), *Phase transitions and critical phenomena* (Vol. 18, pp. 1–142). London: Academic.

Gerlach, W., & Stern, O. (1922). Der experimentelle Nachweis der Richtungsquantelung im Magnetfeld. *Zeitschrift für Physik, 9*, 349–352.

Giustina, M., et al. (2013). Bell violation using entangled photons without the fair-sampling assumption. *Nature, 497*, 227–230.

Goldstein, S. (1998). Quantum theory without observers. *Physics Today,* 42-46 (March 1998)—*Physics Today*, 38-42 (April 1998).

Goldstein, S., & Lebowitz, J. L. (1995). Does the wave function provide a complete description of physical reality? In H. Henry Stroke (Ed.), *The physical review – The first hundred years: A selection of seminal papers and commentaries*. AIP Press.

Gondran, M., & Gondran, A. (2005). Numerical simulation of the double slit interference with ultracold atoms. *American Journal of Physics, 73*, 507–515.

Gondran, M., & Gondran, A. (2016). Replacing the singlet spinor of the EPR-B experiment in the configuration space with two single-particle spinors in physical space. *Foundations of Physics, 46*, 1109–1126.

Grangier, P., Roger, G., & Aspect, A. (1986). Experimental evidence for a photon anticorrelation effect on a beam splitter: A new light on single-photon interferences. *EPL (Europhysics Letters),1*(4).

Grimmett, G. (1999). *Percolation*. Springer.

Handsteiner, J., et al. (2017). Cosmic bell test: Measurement settings from Milky Way stars. *Physical Review Letters, 118*, 060401.

Hänsch, T. W., Schawlow, A. L., & Series, G. W. (1979). The spectrum of atomic hydrogen. *Scientific American*, 94–110.

Hau, L. V., Harris, S. E., Dutton, Z., & Behroozi, C. H. (1999). Light speed reduction to 17 metres per second in an ultracold atomic gas. *Nature, 397*, 594–598.

Hau, L. V., Harris, S. E., Dutton, Z., Behroozi, C. H., Liu, C., Dutton, Z., Behroozi, C. H., & Hau, L. V. (2001). Observation of coherent optical information storage in an atomic medium using halted light pulses. *Nature, 409,* 490–493.

Hensen, B., et al. (2015). Loophole-free Bell inequality violation using electron spins separated by 1.3 kilometres. *Nature, 526,* 682–686.

Henry, J. (1886). *Scientific writings* (Vol. 2). Smithsonian institution.

Hertz, H. (1887). Ueber einen Einfluss des ultravioletten Lichtes auf die electrische Entladung. *Annalen der Physik, 267,* 983–1000.

Johnson, J. B. (1927). Thermal agitation of electricity in conductors. *Nature, 19,* 50–51.

Jönsson, C. (1961). *Zeitschrift für Physik, 161,* 454–474.

Jönsson, C. (1974). Electron diffraction at multiple slits. *American Journal of Physics, 42,* 4–11.

Kamerlingh Onnes, H. (1911). *The resistance of pure mercury at helium temperatures.* Leiden: Comm.

Kamerlingh Onnes, H. (1911). *The disappearance of the resistivity of mercury.* Leiden: Comm.

Kamerlingh Onnes, H. (1911). *On the sudden change in the rate at which the resistance of mercury disappears.* Leiden: Comm.

Kapitza, P. (1938). Viscosity of Liquid Helium below the λ-Point. *Nature, 141,* 74.

Khodabandehlou, F., Maes, C., & Netočný, K., A Nernst heat theorem for nonequilibrium jump processes. arXiv:2207.10313v2 [cond-mat.stat-mech]

Klein, M. J. (1967). Thermodynamics in Einstein's thought. *Science, 157,* 509–516.

Kompaneets, A. S. (1957). The establishment of thermal equilibrium between quanta and electrons. *Soviet Journal of Experimental and Theoretical Physics, 4,* 730–737.

Kragh, H. Preludes to dark energy: Zero-point energy and vacuum speculations. *Archive for History of Exact Sciences.* arXiv:1111.4623 [physics.hist-ph]

Kubo, R. (1986). Brownian motion and nonequilibrium statistical mechanics. *Science, 233,* 330–334.

Kuhn, T. S. (1987). *Black-body theory and the quantum discontinuity, 1894–1912.* University of Chicago Pr. (new Ed. 1987).

Kuhn, T. S. (1957). *The copernican revolution: Planetary astronomy in the development of western thought (Copyright Renewed 1985).* Harvard University Press.

Lamb, W. E., & Retherford, R. C. (1947). Fine structure of the hydrogen atom by a microwave method. *Physical Review, 72,* 241.

Lamoreaux, S. K. (1997). Demonstration of the Casimir force in the 0.6 to 6 μm range. *Physical Review Letters, 78,* 5.

Larsson, J., et al. (2014). Bell-inequality violation with entangled photons, free of the coincidence-time loophole. *Physical Review A, 90,* 032107.

Lummer, O., & Pringsheim, E. (1899). Die Vertheilung der Energie im Spectrum des schwarzen Körpers. *Verhandlungen der Deutsche Physikalische Gesellschaft, 1,* 23–41.

Lummer, O., & Pringsheim, E. (1901). Kritisches zur schwarzen Strahlung. *Annalen der Physik, 6,* 192–210.

Luo, Z., You, Y.-Z., Li, J., Jian, C.-M., Lu, D., Xu, C., Zeng, B., & Laflamme, R. (2019). Quantum simulation of the non-fermi-liquid state of Sachdev-Ye-Kitaev model. *NPJ Quantum Information, 5,* 7.

Lyman, T. (1906). The spectrum of hydrogen in the region of extremely short wave-length. *Memoirs of the American Academy of Arts and Sciences, New Series, 13*(3), 125–146.

Lyman, T. (1914). An extension of the spectrum in the extreme ultra-violet. *Nature, 93*(2323), 241.

Maes, C. (2018). *Non-dissipative effects in nonequilibrium systems.* Springer briefs in complexity.

Maes, C. (2020). Response theory: a trajectory-based approach. In *Frontiers in physics, section interdisciplinary physics.*

Maes, C. (2020). Frenesy: Time-symmetric dynamical activity in nonequilibria. *Physics Reports, 850,* 1–33.

Maes, C. (2020). Fluctuating motion in an active environment. *Physical Review Letters, 125,* 208001.

Maes, C., Meerts, K., & Struyve, W. (2022). Diffraction and interference with run-and-tumble particles. *Physica A, 598*, 127323.

Maiocchi, R. (1990). The case of Brownian motion. *The British Journal for the History of Science, 23*, 257–283.

Maudlin, T. (2015). *Philosophy of physics: Space and time*. Princeton foundations of contemporary philosophy (Vol. 5).

Merli, P. G., Missiroli, G. F., & Pozzi, G. (1976). On the statistical aspect of electron interference phenomena. *American Journal of Physics, 44*, 306.

Michelson, A. A. (1881). The relative motion of the earth and the luminiferous ether. *American Journal of Science, 22*(128), 120–129.

Michelson, A. A., & Morley, E. W. (1886). Influence of motion of the medium on the velocity of light. *American Journal of Science, 31*(185), 377–386.

Michelson, A. A., & Morley, E. W. (1887). On the relative motion of the earth and the luminiferous ether. *American Journal of Science, 34*(203), 333–345.

Millikan, R. A. (1914). A direct determination of "*h*." *Physical Review,4*, 73 (1914)

Miilikan, R. A. (1916). A direct photoelectric determination of Planck's "*h*". *Physical Review, 7*, 362.

Nelson, E. (1967). *Dynamical theoreies of Brownian motion*. Princeton University Press.

Norsen, T. (2005). Einstein's boxes. *American Journal of Physics, 73*, 164–176.

Norton, J. D. (1993). The determination of theory by evidence: The case for quantum discontinuity, 1900–1915. *Synthese, 97*, 1–31.

Nyquist, H. (1928). Thermal agitation of electric charge in conductors. *Physical Review, 32*(1), 110–113.

Oersted, H. C. (1820). Experiments on the effect of a current of electricity on the magnetic needles. Annals of philosophy (Vol. 16, p. 273). London: Baldwin, Craddock, Joy.

Osheroff, D. D., Richardson, R. C., & Lee, D. M. (1972). Evidence for a new phase of solid He^3. *Physical Review Letters, 28*, 885.

Pais, A. (1989). George Uhlenbeck and the discovery of electron spin. *Physics Today*.

Parsegian, A. (2006). *Van der Waals Forces. A handbook for biologists, chemists, engineers, and physicists*. Cambridge University Press.

Paschen, F. (1908). Zur Kenntnis ultraroter Linienspektra. I. (Normalwellenlängen bis 27000 Å.-E.). *Annalen der Physik, 332* (13), 537–570 (1908)

Paschen, F. (1896). Ueber Gesetzmässigkeiten in den Spectren fester Körper. *Annalen der Physik, 294*(7), 455–492.

Pauli, W. (1925). Über den Zusammenhang des Abschlusses der Elektronengruppen im Atom mit der Komplexstruktur der Spektren. *Zeitschrift für Physik, 31*, 765–783.

Penrose, R. (2005). *The road to reality* (p. 632). Alfred A: Knopf.

Perrin, J. (1910). Brownian movement and molecular reality, translated by F. Soddy, London: Taylor and Francis. Reprinted in M.J. Nye, (1986). In *The question of the atom* (pp. 507–601). Los Angeles: Tomash Publishers.

Perrin, J. (1913), *Les atomes*, Paris: F. Alean. Reprinted in D. Li. Hammick (translated) (1990): Atoms, Woodbridge, Connecticut: Ox Bow Press.

Raman, C. V. (1922). *Molecular diffraction of light*. University of Calcutta.

Raman, C. V. (1928). A new radiation.*Indian Journal of physics, 2*, 387–398.

Rauch, H., & Werner, S. A. (2015). *Neutron interferometry: Lessons in experimental quantum mechanics*. Oxford: Wave-Particle Duality and Entanglement.

Rayleigh, L. (1899). On the transmission of light through an atmosphere containing small particles in suspension, and on the origin of the blue of the sky. *The London, Edinburgh, and Dublin Philosophical Magazine and Journal of Science, 47*, 287.

Rayleigh, L. (1918). On the scattering of light by a cloud of similar small particles of any shape and oriented at random. *The London, Edinburgh, and Dublin Philosophical Magazine and Journal of Science, 35*, 209.

Rosenfeld, W., et al. (2017). Event-ready Bell test using entangled atoms simultaneously closing detection and locality loopholes. *Physical Review Letters, 119*(1), 010402.

Rubens, H., & Kurlbaum, F. (1901). Anwendung der Methode der Reststrahlen zur Prüfung des Strahlungsgesetzes. *Annales de Physique, 2,* 649–666.

Russo, L. (2004). *The forgotten revolution. How Science was born in 300 BC and why it had to be reborn.* Berlin: Springer.

Rutherford, E. (1911). The scattering of α and β rays by matter and the structure of the atom. *Philosophical Magazine, 6,* 21.

Rydberg, J. R. (1889). (Récherches sur la constitution des spectres d'émission des éléments chimiques [Investigations of the composition of the emission spectra of chemical elements]. *Kongliga Svenska Vetenskaps-Akademiens Handlingar [Proceedings of the Royal Swedish Academy of Science], 2nd series (in French)23* (11), 1–177.

Sachdev, S. (2011). *Quantum phase transitions* (2nd edn.). Cambridge University Press.

Schmidt-Böcking, H., Schmidt, L., Lüdde, H. J., Trageser, W., & Sauer, T. (2016). The Stern-Gerlach experiment revisited. *European Physical Journal H, 41,* 327–364.

Schmied, R., et al. (2016). Bell correlations in a Bose-Einstein condensate. *Science, 352*(6284), 441–444.

Schrödinger, E. (1936). Probability relations between separated systems. *Mathematical Proceedings of the Cambridge Philosophical Society, 32,* 446–452.

Sengers, J. L. (2002). *How fluids unmix: Discoveries by the school of van der Waals and Kamerlingh Onnes.* Amsterdam: Royal Netherlands Academy of Arts and Sciences.

Shalm, L. K., et al. (2015). Strong loophole-free test of local realism. *Physical Review Letters, 115,* 250402.

Sinha, U., Couteau, C., Jennewein, T., Laflamme, R., & Weihs, G. (2010). Ruling out multi-order interference in quantum mechanics. *Science, 329,* 418–421.

Sorkin, R. (2007). Is the cosmological "constantâŁž a nonlocal quantum residue of discreteness of the causal set type? *AIP Conference Proceedings, 957,* 142–153.

Spohn, H. (1991). *Large scale dynamics of interacting particles.* Berlin: Springer.

Sutherland, W. (1904). The measurement of large molecular masses. In *Report of the 10th meeting of the australasian association for the advancement of science,* Dunedin (pp. 117–121).

Sutherland, W. (1905). A dynamical theory for non-electrolytes and the molecular mass of albumin. *Philosophical Magazine S, 6,* 781–785.

Taylor, G. I. (1909). Interference fringes with feeble light. *Proceedings of the Cambridge Philosophical Society, 15,* 114.

The BIG Bell Test Collaboration. (2018). Challenging local realism with human choices. *Nature,557*(7704), 212–216.

Thomson, G. P. (1927). Diffraction of cathode rays by a thin film. *Nature, 119*(3007), 890.

Tolman, R. C., & Stewart, T. D. (1916). The electromotive force produced by the acceleration of metals. *Physical Review, 8,* 97–116.

Tonomura, A., Endo, J., Matsuda, T., & Kawasaki, T. (1989). Demonstration of single-electron buildup of an interference pattern. *American Journal of Physics, 57,* 117.

Tumulka, R. (2022). *Foundations of quantum mechanics.* Lecture notes in physics (Vol. 1003). Springer.

Tyndall, J. (1869). On the blue colour of the sky: the polarization of skylight, and on the polarization of light by cloudy matter generally. *The London, Edinburgh, and Dublin Philosophical Magazine and Journal of Science, 37,* 384.

Uhlenbeck, G. E., & Goudsmit, S. (1926). Naturwissenschaften *47,* 953 (1925). A subsequent publication by the same authors. *Nature117,* 264 (1926), is followed by a postscript by N. Bohr.

Williams, R. C. (1938). The fine structure of H_α and D_α under varying discharge conditions. *Physical Review, 54,* 558.

Wolf, W. P. (2000). The Ising model and real magnetic materials. *Brazilian Journal of Physics, 30*(4), 794–810.

Wu, M. K., Ashburn, J. R., Torng, C. J., Hor, P. H., Meng, R. L., Gao, L., Huang, Z. J., Wang, Y. Q., & Chu, C. W. (1987). Superconductivity at 93 K in a new mixed-phase Y-Ba-Cu-O compound system at ambient pressure. *Physical Review Letters, 58*, 908.

Wu, X.-L., & Libchaber, A. (2000). Particle diffusion in a quasi-two-dimensional bacterial bath. *Physical Review Letters, 84*, 3017.

Subject Index

Author Index

Printed in the United States
by Baker & Taylor Publisher Services